CONTEMPORARY PORTUGAL

T0251077

For Dominic, Joseph and Luke

Contemporary Portugal
Dimensions of economic and political change

Edited by
STEPHEN SYRETT
School of Social Science, Middlesex University

Routledge
Taylor & Francis Group

LONDON AND NEW YORK

First published 2002 by Ashgate Publishing

Reissued 2018 by Routledge
2 Park Square, Milton Park, Abingdon, Oxon OX14 4RN
711 Third Avenue, New York, NY 10017, USA

Routledge is an imprint of the Taylor & Francis Group, an informa business

Publisher's Note
The publisher has gone to great lengths to ensure the quality of this reprint
but points out that some imperfections in the original copies may be
apparent.

Disclaimer
The publisher has made every effort to trace copyright holders and welcomes
correspondence from those they have been unable to contact.

A Library of Congress record exists under LC control number: 2002024892

ISBN 13: 978-1-138-71729-9 (hbk)
ISBN 13: 978-1-138-71727-5 (pbk)
ISBN 13: 978-1-315-19641-1 (ebk)

Contents

List of Figures

List of Tables and Boxes

List of Contributors

David Corkill is Professor in Iberian Studies at Manchester Metropolitan University. He has written widely on Portugal's economy and politics. He has published the *Portuguese Economy Since 1974* (Edinburgh University Press, 1993) and numerous articles including 'Multiple identities, racism and immigration in Spain and Portugal' in B. Jenkins and S. Sofos (eds) *Nation and Identity in Contemporary Europe* (Routledge, 1996) and 'The Iberian economies: divergence to convergence?' in B. Foley (ed) *European Economies Since the Second World War* (Macmillan, 1998). His most recent publication was *The Development of the Portuguese Economy: a Case of Europeanization* (Routledge, 1999) and he is currently working on a new book, *The Economies of Spain and Portugal: a Regional Perspective.*

Martin Eaton is Lecturer in European Regional Development at the University of Ulster (Coleraine). He has published widely in French, German and British journals in the area of Portuguese human geography. His research has centred on the topics of Portuguese regional development, industrial geography, labour market problems, child workers, and his current area of interest, immigration.

Richard A.H. Robinson is Reader in Iberian History at the University of Birmingham, where he has taught modem history since 1965. He has written extensively on Portuguese and Spanish history in the Twentieth Century. He has been a *bolseiro* of the Fundação Calouste Gulbenkian. His better-known publications include *Contemporary Portugal: A History (1979), The Origins of Franco's Spain (1970)* and *Fascism: The International Phenomenon (1995).*

Carlos Nunes Silva is Professor Auxiliar in the Department of Geography at the University of Lisbon where he completed his PhD in regional and local planning in 1995. His area of specialist research is that of local government policies, however he has interests in urban and regional planning, public administration and political geography. He has published widely on these topics in both Portuguese and English. He published *Política Urbana em Lisboa, 1926-74* (Livros Horizonte, 1994) and recent English language publications include: 'Local government, ethnicity and social exclusion in Portugal' in A. Khakee et al. (eds) *Urban Renewal, Ethnicity and Social Exclusion in Europe* (Avebury, 1999); 'Local finance in Portugal', *Environment & Planning,* (1999: 16); and (with Syrett) 'Regional Development Agencies in Portugal', *Regional Studies* (2001: 35(2)).

Carlos Pereira da Silva is Lecturer in the Department of Geography and Regional Planning at the Universidade Nova de Lisboa and a researcher at the Centre of Geographical Studies and Regional Planning. His main research interest is coastal management, but other interests include the management of protected areas and landscape perception. He is currently completing his PhD project 'The management of the Alentejo coast: tourist demand as constraint'. He has written widely on issues of coastal development and protected natural areas and is currently involved in a research project on the 'Assessment of Perception Patterns in Coastal Areas with Tourist Potential'.

Helena Sousa is Director of the Media Programme and Deputy Head of the Communication Sciences Department at the Universidade do Minho. Formerly she worked as a journalist on the Portuguese national newspaper *Jornal de Notícias* and in the Journal of Commerce in New York. She completed her PhD on 'Communications policy in Portugal and its links with the European Union: an analysis of the telecommunications and television broadcasting sectors' at the City University in London. Her principal research interests include communications policy and regulation, international broadcasting and journalistic sources.

Stephen Syrett is Principal Lecturer in Economic Geography at Middlesex University. He has been actively researching processes of socio-economic restructuring, local and regional economic development and changing structures of local and regional governance within Portugal for over a decade. He has published widely on these topics including *Local Development: Restructuring, Locality and Economic Initiative in Portugal* (Avebury, 1995) and numerous journal articles including 'The politics of partnership' in *European Urban and Regional Studies*, (1997: 4(2)); 'International regulation and regional development', in *European Planning Studies*, (1996: 6(4)) and (with Silva C.N.) 'Regional Development Agencies in Portugal', *Regional Studies* (2001: 35(2)).

Allan M. Williams is Professor of Human Geography and European Studies in the University of Exeter. He has a general interest in questions of European economic development and European integration, with a particular focus on the Mediterranean region. He has been exploring some of the social and economic issues related to uneven development in Portugal since the late 1970s and has published on tourism, migration, small firm formation, housing policies and the effects of integration. He has recently completed a major investigation into the causes and consequences of international retirement migration to southern Europe, including a case study of the Algarve. His publications include (co-editor) *Divided Europe: Society and Territory* (Sage, 1999), *Tourism and Economic Development: European Experiences* (Wiley, 1998), *European Tourism: Regions, Spaces and Restructuring* (Wiley, 1995) and *Turkey and Europe* (Frances Pinter, 1993). He is also co-editor of the journal, *European Urban and Regional Studies*.

Preface

Portugal experienced a period of rapid and dramatic change over the final decades of the twentieth century that impacted upon all aspects of economic, political and social life. After the instability that characterised the aftermath of the 1974 revolution, the 1980s and 1990s provided a period of unprecedented political stability and economic modernisation, during which Portugal converged rapidly with the wealthier member states of the European Union. This volume takes this recent period as its focus. The contributions in this volume seek to provide an accessible overview of key dimensions of economic and political change and identify the tensions and policy challenges that such rapid change produced. In so doing they reveal something of the complexity of contemporary Portugal; an outward looking modern, democratic and European state, but one where the legacy of its recent traditional, colonial and often inward looking past, continues to influence and shape its development in the twenty-first century.

This edited collection has its roots in a one day conference entitled *Portugal at the Millennium: Economic, Social and Political Change in Contemporary Portugal*, which took place at Canning House, London. This conference was organised by the School of Social Science of Middlesex University in conjunction with Canning House and the Anglo-Portuguese Society and the interest this generated provided the impetus for the current volume. I would like to thank these organisations, and in particular Ann Waterfall, Julian Amey, Steve Chilton and Josie Joyce for their help in organising this event. In editing this collection I would like to thank all of the contributors, but owe a particular debt to Martin Eaton and David Corkill for their guidance and support throughout the process of completing this book. I would also like to thank Carlos Nunes Silva and Rui Jacinto for their continued research support and insights into contemporary Portugal.

My thanks go to the School of Social Science at Middlesex University for their generous support of this project. In the preparation of the text and figures. I would like to thank the staff of the Technical Unit, particularly Paul Joyce for his initial work, and Natalie Tuckwell for her professionalism and unceasing enthusiasm and commitment during a number of frustrating delays. And never least, I would like to thank Claire, Joseph and Luke for everything.

List of Abbreviations

AACS	*Alta Autoridade para a Comunicação Social* (High Authority for the Media)
AD	*Aliança Democrática* (Democratic Alliance)
ADIM	*Associação Democrática Independente de Macau* (Independent Democratic Association of Macau)
AM	*Assembleia Municipal* (Municipal Assembly)
AML	*Área Metropolitana de Lisboa* (Metropolitan Area of Lisbon)
ANMP	*Associação Nacional dos Municípios Portugueses* (National Association of Portuguese Municipalities)
APU	*Aliança 'Povo Unido'* (United Peoples Alliance)
BE	*Bloco de Esquerda* (Left Bloc)
BSCH	Banco Santander Central Hispano
CAP	Common Agricultural Policy
CCR	*Comissão de Coordenação Regional* (Regional Planning Commission)
CDS	*Centro Democrático Social* (Democratic Social Centre)
CDU	*Coligação Democrática Unitária* (Democratic Unity Coalition)
CEDITC	*Comissão para o Estudo do Desenvolvimento Institucional e Tecnológico das Comunicações* (Commission for the Study of the Institutional and Technological Development of Communications)
CEFAMOL	*Associação dos Industriais de Moldes* (Association for the Moulds Industry)
CENFIM	*Centro de Formação Profissional da Indústria Metalúrgica e Metalomecânica*

	(Training Centre for the Metallurgical and Metal Mechanics Industry)
CENTINFE	*Centro Tecnológico para a Indústria de Moldes e Ferramentas Especiais* (Technological Centre for the Moulds and Specialised Tools Industry)
CM	*Câmara Municipal* (Municipal Chamber)
CN	*Comunicações Nacionais* (National Communications – financial holding company)
CNA	*Comissão Nacional do Ambiente* (National Commission for the Environment)
CPLP	*Comunidade dos Países de Lingua Portuguesa* (Community of Portuguese-speaking Countries)
CSF	Community Support Framework
CTT	*Correios e Telecomunicações de Portugal* (Post and Telecommunciations of Portugal)
DGAA	*Direcção-Geral de Administração Autárquica* (Directorate General for Local Authority Administration)
DVB-T	Terrestrial Digital Video Broadcasting
EC	European Commission
EEC/EC	European Economic Community/European Community
EFTA	European Free Trade Association
EMU	European Monetary Union
ER	Emigrant Remittances
ERM	Exchange Rate Mechanism
ETAR	*Estação de Tratamento de Águas Residuais* (Sewage Treatment Station)
EU	European Union
FCM	*Fundo de Coesão dos Municípios* (Municipal Cohesion Fund)
FDI	Foreign Direct Investment
FGM	*Fundo Geral dos Municípios* (General Municipal Fund)
FRS	*Frente Republicano e Socialista* (Republican and Socialist Front)
GAT	*Gabinetes de Apoio Técnico* (Technical Support Units)
GATT	General Agreement on Tariffs and Trade

GDP	Gross Domestic Product
ICEP	*Investimentos Comércio e Turismo de Portugal* (Portuguese Investments, Trade and Tourism)
ICP	*Instituto das Comunicações de Portugal* (Portuguese Communications Institute)
ICS	*Instituto da Comunicação Social* (Media Institute)
IMF	International Monetary Fund
INE	*Instituto Nacional de Estatística* (National Institute for Statistics)
IRS	*Imposto Sobre o Rendimento das Pessoas Singulares* (Personal Income Tax)
ISCTE	*Instituto Superior de Ciências do Trabalho e da Empresa* Higher Institute for the Sciences of Work and the Firm
IT	Information Technology
LBA	*Lei de Bases do Ambiente* (Structural Law for the Environment)
LPN	*Liga para a Protecção da Natureza* (League for the Protection of Nature)
LRF	Legally Resident Foreigners
MAOT	*Minstério do Ambiente e Ordenamento do Território* (Ministry of Environment and Territorial Management)
MDP	*Movimento Democrático Português* (Portuguese Democratic Movement)
MEPAT	*Ministério do Equipamento, do Planeamento e da Administração do Território* (Ministry of Equipment, Planning and Territorial Management)
MERCOSUR	Southern Cone Common Market (Latin America)
MFA	*Movimento das Forças Armadas* (Armed Forces Movement)
MOPTC	*Ministério das Obras Públicas, Transportes e Comunicações* (Ministry of Public Works, Transport and Communications)
MNC	Multinational Corporation
MPAT	*Ministério do Planeamento* Ministry of Planning and Administration of the Territory
NAFTA	North American Free Trade Agreement
NGO	Non Governmental Organisation

NIEs	Newly Industrialising Economies
OPEC	Organisation of Petroleum Exporting Countries
PALOPs	*Países Africanos de Lingua Oficial Portuguesa*
	Portuguese speaking African Countries
	(Angola, Mozambique, Guinea-Bissau, Cape Verde,
	S. Tomé and Príncipe).
PCP	*Partido Comunista Português*
	(Portuguese Communist Party)
PC(R)	*Partido Comunista (Reconstituido)*
	((Reconstituted) Communist Party)
PDM	*Plano de Director Municipal*
	(Municipal Lead Plan)
PEDAP	*Programa Específico de Desenvolvimento da Agricultura Portuguesa*
	(Specific Programme for the Development of Portuguese Agriculture)
PEDIP	*Programa Específico de Desenvolvimento da Indústria Portuguesa*
	(Specific Programme for the Development of Portuguese Industry)
PEV	*Partido Ecologista 'Os Verdes'*
	(Ecology Party 'the Greens)
POOC	*Plano de Ordenamento da Orla Costeira*
	(Plan for the Management of the Coastal Margin)
PP	*Partido Popular*
	(Popular Party)
PPD	*Partido Popular Democrático*
	(People's Democratic Party)
PRD	*Partido Renovador Democrático*
	(Democratic Renewal Party)
PS	*Partido Socialista*
	(Socialist Party)
PSD	*Partido Social Democrata*
	(Social Democrat Party)
PSN	*Partido de Solidariedade Nacional*
	(National Solidarity Party)
PSR	*Partido Socialista Revolucionário*
	(Revolutionary Socialist Party)

R&D	Research and Development
RDP	*Rádiodifusão Portuguesa*
	(State owned national radio station)
RR	*Rádio Renascença*
	(Catholic owned national radio station)
RTP	*Rádiotelevisão Portuguesa*
	(State owned national television company)
SADAC	Southern African Development Community
SEA	*Secretaria de Estado do Ambiente*
	(Secretary of State for the Environment)
SEARN	*Secretaria de Estado do Ambiente e Recursos Naturais*
	(Secretary of State for the Environment and Natural Resources)
SEF	*Serviço de Estrangeiros e Fronteiras*
	(Foreigners and Frontiers Service)
SEM	Single European Market
SETC	*Secretaria de Estado dos Transportes e Comunicações*
	(Secretary of State for Transport and Communications)
SIC	*Sociedade Independente de Comunicação*
	(Independent television channel)
SMEs	Small and Medium Size Enterprises
TDP	*Teledifusora de Portugal*
	(Television distribution company)
TLP	*Telefones de Lisboa e Porto*
	(Telephones of Lisbon and Porto)
TP	*Telecom Portugal*
TVI	*Televisão Independente*
	(Independent television channel associated with the Catholic Church)
UDP	*União Democrática Popular*
	(People's Democratic Union)

1 Portugal Transformed

STEPHEN SYRETT

Introduction

That Portugal underwent a period of dramatic societal change in the last 25 years of the twentieth century is beyond dispute. One need look no further than the new landscape of motorways, apartment blocks, hypermarkets, advertising hoardings and mobile phones, to know that change has been profound and rapid. Portugal increasingly conforms to the model of an advanced western European state in terms of its economic structure, democratic political tradition, models of social organisation and state provision of welfare services, as well as in the aspirations and expectations of its population. Socio-economic indicators confirm increasing convergence to European norms whether in terms of diet, electoral turn out or demographic characteristics.

The pace of modernisation has been particularly impressive. Fundamental changes that in other European countries took place across half a century or more have been compressed into little more than two decades. Yet the pace and scale of change has brought with it fragility and vulnerability. There remain doubts as to the strength of the foundations of the new social, political and economic systems and whether they can be sustained and built upon in the future. Significant gaps between Portugal and its advanced European partners remain whether in terms of the strength and diversity of its economic base, levels of education and training, or the quality and efficiency of public services. Continuity is also very much in evidence. Dominant economic and political elites have survived and reinvented themselves to remain at the heart of modern Portugal. Rapid change has not swept away traditional lifestyles; rather the contemporary landscape is marked by the juxtaposition of traditional and modern forms of production, consumption and social reproduction.

The contradictions and discontinuities that characterise a period of rapid transition are defining features of contemporary Portugal. As Portugal seeks to establish a new place and identity within Europe, and negotiate the uneasy transition towards convergence with the economic and political structures characteristic of the modern democratic states of northern Europe, it is confronted by an array of tensions and paradoxes rooted within unfolding processes of change. This volume explores the economic and political dimensions of Portugal's current journey 'backwards out of the big world' (Hyland, 1997) seeking to understand the nature of such change and identify the new challenges it has created.

The roots of many of the processes of contemporary change can be found in the 1960s (Barreto, 1996a). In this period mass out-migration from interior rural areas resulted in rural depopulation and the growth of urbanisation processes centred on the Lisbon and Porto metropolitan areas. A new and deeper integration into the international economy also emerged via labour migration to northern Europe and the development of manufacturing and tourist industries. The social functions of the state began to develop in areas such as education and social security, whilst culturally the arrival of television and social mixing arising from the experiences of the colonial wars were important social factors. The shift in orientation towards Europe through economic integration and away from Africa, marked the start in a major repositioning in Portuguese outlook from Atlanticist imperial power to small European nation state.

The revolution of 1974 marked a watershed in the shift to a new era. The resultant 1976 constitution established universal suffrage in national and local elections for the first time in Portugal's political history. The major historical changes of this period - the establishment of democracy, decolonisation, the beginnings of European integration - are analysed in a number of English language texts (Baklanoff, 1978; Robinson, 1979; Graham and Makler, 1979; Gallagher, 1983; Graham and Wheeler, 1983; Sousa Ferreira & Opello, 1985). Yet since this period the story has moved on (Barreto, 1996b; Santos, 1993), and in a manner that few predicted. This volume focuses upon the post-revolutionary period but with particular emphasis upon change in the 1980s and 1990s. In this period Portugal has enjoyed a degree of economic and political stability and has been driven by a vision of European integration that has enabled the playing out of rapid and dramatic change. The starting point for this analysis is not therefore that of a country with its 'head in the First world and its feet in the Third', but rather a country with a social, economic and political profile convergent with the established economies and democracies of western Europe; a convergence largely unthinkable in 1974 and the turbulent period that followed.

Portugal's newly acquired 'maturity' means it now faces many of the same dilemmas as advanced western nations (falling electoral turnouts, immigration, drug abuse, rising criminality etc.[1]). However the constitution of these new problems, alongside more traditional ones, must be understood within the particular context of the nature and scope of the recent transition process within Portugal. This process of change is of wider significance too. In the European context many of the smaller transitional economies that hope to join the European Union over the forthcoming decade are interested by the apparent economic and political success of Portugal. Analysis of the Portuguese experience of rapid transition thus provides insights into the challenges of European integration that may be informative to other applicant countries. Although it is beyond the scope of this volume to analyse comprehensively all aspects of contemporary economic and political change in Portugal, the contributions in this book provide a starting point for a fuller understanding of a number of key dimensions of the transition process.

Towards Convergence?

To date, the apex of the process towards European convergence was Portugal's presence as one of the founding members of the first phase of economic and monetary union in 1999. The Portuguese 'economic miracle' talked of by some commentators reflected the strong macro-economic performance of the Portuguese economy across the 1980s and 1990s (see Table 1.1; see also Chapters Two and Three). During this period consistently above average GDP growth rates enabled Portuguese GDP per capita to rise from 52.8 per cent of the EU average in 1985 to 73.3 per cent in 1999. Across this same period the inflation rate fell from 19.4 per cent to 2.3 per cent. Marked improvements in public finance included a reduction in the public account deficit from 6.1 per cent of GDP in 1993 to 2.3 per cent and a reduction in public debt; a performance that ensured Portugal met the convergence criteria for public finance laid down in the Maastricht Treaty. Yet despite this strong economic progress evidence of structural differences remain. Continued low levels of productivity (still only 66 per cent of EU average), the domination of the business structure by small and medium size enterprises (SMEs), low levels of investment in research and development (R&D), and specialisation in traditional low technology industrial products and processes (see Chapter Two); all indicate that the Portuguese economy remains underdeveloped in certain fundamental components compared to Europe's established industrial economies.

The process of convergence towards European norms is a consistent feature across a range of socio-economic indicators. With respect to population characteristics, birth rates have fallen dramatically from 24 per thousand in 1960 to 11.4 in 1997, a figure in line with the EU average of 10.8 per thousand. The age structure is also now similar to elsewhere in the EU with the population of 14 and under representing 17.6 per cent of total population, and the over 65 population, 14.7 per cent. Although the percentage of elderly remains less than the EU average (15.6 per cent), the trend towards an ageing population is one of the most rapid in Europe, and in certain regions (e.g. Alentejo, Algarve and the Centre) levels are already above the EU average.

The structure of the labour market increasingly resembles that of the more advanced EU economies. An increased number employed within

Table 1.1 Main Macroeconomic Indicators, 1985-99

	(%)			
	1985	**1960**	**1995**	**1999**
Per capita GDP	52.8	60.7	70.6	73.3
(compared with EU average) (a) (c)				
Productivity	49.3	55.7	63.8	65.8
(compared with EU average) (a) (b)				
Rate of unemployment	8.7	4.6	7.3	4.6
Rate of inflation	19.4	12.4	4.5	2.3
General government balance	-10.5	-5.1	-5.7	-1.9
(as % of GDP) (c)				
Gross general government debt	60.8	64.2	64.7	56.1
(as % of GDP) (c)				
Current account balance	0.4	-1.7	-5.1	-2.5
(as % of GDP)				
Degree of openness of the economy (d)	35.8	37.6	34.3	35.4
Rate of investment (as % of GDP)	23.9	27.6	23.6	25.7

Source: European Commission
(a) In purchasing power standards (PPS)
(b) GDP per person in employment; the EUR 15 averages for 1985 and 1990 do not include the GDR
(c) Calculated in accordance with the definitions applicable under the excessive deficit procedure
(d) (Exports + Imports) / 2 / GDP * 100 (at current prices)

tertiary activities (over 50 per cent), a stable secondary sector (35 per cent) and a declining, albeit still large, agricultural sector (13 per cent) mark a fundamental shift from the rurally dominated economy of the early 1960s (see Table 1.2 and Chapter Three). These changes have drastically restructured some of the traditional economic activities strongly associated with Portugal (e.g. fisheries,[2] wine production, textiles and clothing) and witnessed the emergence of important new sectors of economic activity (e.g. producer services, telecommunications, car components). However the labour market remains distinctive in key dimensions. Unemployment has always been, and remains, low by European standards, averaging 4.0 per cent in the late 1990s with peaks in 1985 and 1996 of 8.6 per cent and 7.3 per cent respectively. Such low levels of unemployment reflect the particular constitution of the labour market; one characterised by high flexibility but low levels of skills and productivity. It is also a labour market that exhibits high and growing levels of female participation, from 15 per cent of women working formally in 1960 to 49.4 per cent in 1997, a figure above the EU average of 45.6 per cent.

In the provision of basic social services recent decades have seen major advances towards EU norms. In health provision, the rate of expenditure rose to 8.2 per cent of GDP in 1997 (compared to 8.6 per cent EU average), a rapid increase from 5.0 per cent in 1994, and the lowly figure of only 1.0 per cent in 1960. As a result by the late 1990s levels of health services had grown significantly. In the 15 years after the revolution there was a tripling in the number of doctors, and by 1997 there were three doctors per thousand inhabitants compared to less than one per thousand in 1960. Similar increases were achieved in the numbers of nurses and hospital beds although significant regional differences in the levels of health care provision remain (Ministério do Planeamento, 2000). Rates of infant mortality fell dramatically to 6.4 per thousand live births in 1997. However infant and general mortality rates still remain above EU averages and life expectancy is below the European average, -2.7 years for men and -1.8 years for women.

The area of education witnessed similar major advances from very low base levels. In 1960 education spending accounted for only 1.5 per cent of GDP reflecting the low priority of mass education during the *Estado Novo* (1926/33-74). Spending levels rose to 3.8 per cent (of GDP) in 1975 and to 5.5 per cent by 1992; levels of investment that underwrote a major expansion in the numbers receiving primary, secondary and higher education (from 1,140 000 in 1960 to 2,290 000 in 1995). Despite this, by European standards

Table 1.2 Portugal: Socio-Economic Characteristics

	Year	Unit	
Demographic and Social Indicators			
Area		100km^2	919
Population	1997	10^3	9 957.3
Population density	"	inhab./km^2	108.3
Birth rate	1997	‰	11.4
Death rate	"	"	10.5
Ageing index	"	%	88.5
Health			
Hospitals	1997	No	215
Doctors per 1 000 inhabitants	"	‰	3.0
Beds per 1 000 inhabitants	"	"	4.0
Education			
Number of pupils			
Primary	1998/99	No	1 158 794[1]
Secondary	"	"	381 118[1]
Higher	"	"	N/a
Educational establishments			
Primary	1995/96	No	12 874
Secondary	"	"	664
Higher	"	"	290
Net primary school entolment rate	1996/97	%	82.1[1]
Level of scholastic education in	1998	%	80%≤9thyr
age strata 25-64 years			10%=12thyr
			10%>12thyr
Culture and Recreation			
Newspapers and magazines	1995	10^3	522 670
– yearly circulation			
Museums	"	No	341
Libraries	"	"	1 614
Public entertainment - performances	"	"	150 645
Economic Indicators			
GVA pm	1996	ESC10^6	15 368 681
Primary sector	"	%	4.1
Secondary sector	"	"	33.9
Tertiary sector	"	"	61.9
GDP/inhab.	1997	ESC10^3	1 797
Disposable per capita family income	1995	"	1 137
Poverty rate [3]	1995	%	22.7[1]
Employment	1998	10^3	4 738.8
Primary sector	"	%	13.5
Secondary sector	"	"	35.8
Tertiary sector	"	"	50.7
Activity rate	"	%	50.0

Table 1.2 (cont.)

	Year	Unit	
Economic Indicators (cont.)			
Unemployment rate	1998	"	5.0
Female unemployment rate	"	"	6.3
Youth unemployment rate	"	"	10.2
Long-term unemployment[4]	"	"	42.6
Environment[2]			
Water supply	1997	%	86[1]
Waste water drainage	"	"	68[1]
Urban waste water treatment	"	"	40[1]
Treatment of solid urban waste	"	"	24[1]
Comfort Indicators			
Dwellings with			
Internal running water	1997	%	93.2
Fixed installations - Bath/shower	"	"	89.1
Telephone	"	"	79.7
Transport and Communication			
Road network	1996	Km	9 752[1]
Main routes	"	"	2 558[1]
Supplementary routes	"	"	2 416[1]

Source: Ministério do Planeamento, 2000, p.35
(1)   Mainland
(2)   Overall level of provision to population
(3)   Percentage of population below poverty threshold (population with income below 50% of national average income/total population)
(4)   Unemployment for more than 1 year

educational attainment remains weak. Illiteracy levels remain high at 10.4 per cent of the population in 1996. For 80 per cent of the population primary level education (nine years of schooling) is the highest qualification (compared to an EU average of 46 per cent), and only 10 per cent have a secondary education diploma (EU average of 41 per cent).

Major investments in basic infrastructures have resulted in a massive upgrading of the road and rail networks. Environmentally, substantial improvements from very low pre-existing levels were achieved in basic sanitation and domestic and industrial waste disposal. Yet at the same time, rapid economic modernisation also increased problems of pollution and environmental degradation (see Chapter Seven). In terms of indicators of well-being, the 1970s and 1980s witnessed notable improvements with regard to basic domestic supply infrastructures (e.g. electricity, public drainage systems, indoor tap water, toilets and bathrooms), again from very low base levels. For example in 1960 only 29 per cent of households had a domestic water supply, a figure which rose to 82 per cent in 1991 and 90 per cent in 1999, whilst households served with electricity rose from 41 per cent in 1960 to 90 per cent in 1991.

As the level of basic domestic infrastructures improved, from the mid 1980s onwards dramatic increases were evident in second order indicators of well-being, such as the number of households with televisions, heating, telephones, freezers, washing machines and cars. In this two indicators are particularly noteworthy, both as a reflection of improvements in the standard of living but also of important changes in the nature of contemporary Portuguese society. First, by 1994, 96 per cent of households had televisions. Second, the 1990s witnessed a mobile phone revolution which saw growth from 1 per cent of the population owning mobile phones in 1991, to 52 per cent in 2000, one of the highest levels of mobile phone use within Europe. Such changes indicate how the isolationism characteristic of many regions in the 1960s and 1970s has been transformed by the adoption of mass and personal communication technologies. However these rapid increases in consumer items have come at a price. Fuelled by low interest rates, the 1990s witnessed an unprecedented consumer boom that saw the Portuguese shift from a nation of savers to a nation of borrowers. As a result, by 2001 personal debt had risen to 100 per cent of disposable income compared to a level of only 18.5 per cent in 1990 (Financial Times, 2001).

Despite the evidence of impressive improvements across a range of socio-economic indicators, an enduring feature of Portugal's rapid socio-economic modernisation has been the persistence of poverty and social inequality. The distribution of income within Portugal is the most unequal within the EU (Ministério do Planeamento, 2000). This reflects high levels of poverty, the highest within the EU when defined as the percentage of families or individuals with an equivalent average monthly income below the poverty line. Poverty and social exclusion is concentrated among those who endure a precarious existence, often on the margins of the formal economy. These include those in low paid employment, a category which includes a significant number of underemployed workers (e.g. in agriculture), those operating informally (e.g. immigrant workers and child workers),[3] as well as those existing on meagre levels of social security payments, notably the unemployed and the elderly in receipt of state pensions. Social problems have been exacerbated further by the undermining of traditional social ties and norms as a result of rapid processes of modernisation and urbanisation. The erosion of customary forms of social capital has weakened the basic family and community support structures that marginalised and low-income communities have habitually drawn upon.

Inequality is also evident across other dimensions. Despite the high level of workforce participation and the rising level of educational achievement of women (women now account for 64 per cent of university graduates), gender inequality is evident with respect to wages, with women's wages representing only 71.7 per cent of corresponding men's wages in 1995. Inequality is also evident with respect to the spatial distribution of wealth and services. The expansion of market systems and national level provision of public services has reduced some aspects of the profound regional inequalities between marginalised rural interior regions and the coastal metropolitan areas evident in the 1960s. However, the continued growth of the major metropolitan areas in the coastal belt and loss of population from interior rural areas, continues to result in major regional wealth differentials, with Lisbon and the Tagus valley the only region approaching EU average levels of GDP (see Chapter Three). However recent socio-economic change has also led to rising inequalities between localities, with the growth of pockets of deprivation within the expanding metropolitan areas, easily visible in the unplanned and poorly serviced shanty towns and *barracos* which characterise many suburban areas of the larger cities and towns.

The Competitive Basis of Portugal

Taken together, the recent evolution of socio-economic indicators tells a common story; rapid progress from very low levels across a short period of time, yet by EU standards still significant short falls in many fundamental aspects. A useful snapshot of the state of Portugal's social, economic and political development profile on the eve of the twenty-first century is provided by the 'SWOT analysis' contained in the proposed Third Community Support Framework for the 2000-2006 period (Ministério do Planeamento, 2000; see Table 1.3). The inclusion of this descriptive analysis of Portugal's strengths, weaknesses, opportunities and threats, reflects the acceptance by the Portuguese administrative elite of the prevailing business oriented approach to national development; one which brings the language and tools of business management to the management of national economies.[4]

Table 1.3 SWOT Analysis of Portugal, 1999

Strengths	Weaknesses
• Stable and well established democratic political system • Open economy and strong relations with European markets • Rapid convergence with EU and founder member state of EMU • Improved levels of attendance at all levels of education, particularly at pre-school and higher levels • Flexible labour market with high rates of female participation • Strong array of social institutions and voluntary organisations available to combat social exclusion in collaboration with the state • Privatisation and liberalisation processes which permitted the modernisation of the financial sector, infrastructures, distribution and certain industrial sectors • Diverse, dynamic and increasingly internationalised SME sector • Strong direct investment via large firms in manufacturing, finance and infrastructures, and internationally in the distribution sector • Strong integration into international distribution networks and patterns of European consumption • Rapid modernisation of telecommunications/audio-visual network • Involvement of major businesses from financial services and distribution in the development of e-commerce • Presence of diverse professional, scientific and artistic skills, particularly within major metropolitan centres, appropriate to the development of higher level service and technological industries • Presence of medium sized towns in dynamic economic regions capable of providing a more balanced urban network • Some improvements in housing supply via owner-occupier housing boom • Substantial improvement in health service infrastructures	• Inefficient operation in basic areas of law, justice and public administration • Lack of reform of social provisions of the state in areas such as health, social security and taxation, needed to improve financial sustainability, coverage and efficiency • Peripherality of geographical position • Industrial structure, often geographically concentrated, with strong exposure to competition from developing countries • Low levels of qualifications in work force and problems of structural unemployment • Significant levels of school failure and poor performance in areas such as mathematics and science • Educational and training systems poorly articulated to current and future skill demands • Lack of basic knowledge in information and communication technologies and insufficient training of young people in these and associated areas • Low productivity associated with low levels of training and education, weak business organisation, and low value products • Weak development of stock markets • Financial system poorly structured to support innovation and R&D activities • Weak ability to attract international investment into leading edge manufacturing and high value service industries • Problematic urban system characterised by weak planning in major metropolitan areas, diffuse urbanisation in the North and Centre coastal area, and lack of co-operation between urban centres • Existence of social exclusion and poverty

Table 1.3 (cont.)

Opportunities	Threats
• Membership of EU	• Opening of European Single Market to competition from developing countries via the development of EU trade policy
• Geographical position favourable to major sea transportation routes and air links between Europe and other continents	• Phase of slow growth in European economy given strong dependence on European markets
• Climate, environment and culture favourable to the development of tourism and retirement migration	• Domestic problems in achieving strong growth in productivity and GDP liable to create tensions between levels of competitiveness and employment
• Strong links to dynamic localities in China and Asia as well as traditional links with Latin America and Africa	• Competitive pressures on small subsistence agricultural producers in the North and Centre and on traditional industrial sectors where labour force retains strong ties with agriculture
• Development of cyberspace activities to overcome problems of geographical peripherality within Europe	• Vulnerability of recently developed industrial clusters (e.g. cars) due to investment decisions of large multinational companies
• Existence of decentralised system of further and higher education and research establishments available to support the modernisation of industry	• Emigration of high skills executives and professionals due to lack of domestic opportunities
• Availability of skills in computing, IT, biology, health and biotechnology, for future development of technological activities	• Concentration of strategic operations serving the Iberian peninsula within Spain
• Existence of road network capable of enabling international access and domestic cohesion	• Poorly planned over-exploitation of environmentally sensitive and economically valuable areas
• Programme of major investment in water supply, basic sanitation, collection and treatment of waste, to improve environmental quality	• Inability of metropolitan areas of Lisbon and Porto to act as motors for transforming the profile of national economic activities and international links
• Programme of investment in water supply and agriculture to permit the development of irrigated farming and agrifood activities	• Difficulties in strengthening interior medium sized towns, agri-environmental activities and competitive centres of agricultural production to prevent rural desertification
• Potential to increase forested areas and to further develop timber, paper, cellulose, packaging and cellulose fibre industries	

Source: Translated and abridged from Ministério do Planeamento, 2000, pp.32-34

Although the precise details of this particular piece of descriptive analysis are open to critical debate, the story it tells is an intriguing one (see Table 1.3). The language utilised clearly demonstrates a desire to position Portugal as an advanced, internationalised, open and outward looking European economy, convergent and comfortable with European norms of production, distribution, consumption and social welfare. This was of course the narrative that the European Commission wanted to hear, and in this respect the analysis reflected the political process of positioning for Community funds. However it is also an account of contemporary Portugal around which a surprisingly strong, and largely uncontested, national consensus developed across the 1980s and 1990s.

The analysis identifies Portugal's principal strengths as an open, liberalised and flexible economy, underpinned by a stable democracy, which exhibits dynamism in selected sectors, firms and regions. A presence in some leading edge sectors is evidenced through rapid advances in telecommunications, information technologies and increasing skill levels in computer technologies, with strong spatial concentration of the more advanced manufacturing and service sectors in Lisbon and its wider region. Whilst strengths are seen as primarily economic, weaknesses tend to focus on the ongoing failure of social systems and public administration to deliver sufficiently high levels of education and health and tackle problems of social exclusion, poverty and rapid unplanned physical development. However, economic weaknesses are also noted in relation to the traditional industrial structure, low levels of productivity and the underdeveloped nature of stock markets and financial systems.

Opportunities for future development recognise the unique opportunity provided by a further period of substantial EU support (between 2000-2006) for further rapid improvements in basic infrastructures, roads, the environment and industrial modernisation. Portugal's competitive advantages in natural resources, climate and environment for tourism and retirement migration, and forestry, are also noted. Although Portugal's peripheral geographical position within Europe is seen as a weakness, its potential role as a European link into Africa, Latin America and Asia is seen as a potential advantage. The possibility of developing existing pockets of advanced skills in technological areas is seen as another potential opportunity to lead the Portuguese economy towards a more advanced industrial structure.

Although Portugal's open, liberalised and internationally oriented economy is seen as major strength, it is also evident that the majority of perceived threats also directly relate to this position. Fears over loss of investment and highly skilled workers, reliance on the dynamism of the European economy, lack of environmental regulation, and the inability of small firms in traditional sectors to compete with developing economies, all testify to the evident vulnerability of a small, open economy in an increasingly competitive global economy. Such an analysis raises important questions. Can Portugal retain and enhance its competitive position internationally? Or will it lose out to other developing regions within the global economy, and other EU member states (e.g. Spain) within the European arena? Can it develop the necessary level and quality of social and political systems to underpin and sustain an advanced industrial economy?

The subsequent chapters of this book provide insights into these broad questions through analysis of selected dimensions of recent economic and political change. A number of themes run through these chapters but the interrelated notions of economic globalisation and the reconfiguration of the central state appear particularly important to understanding processes of contemporary change.

Portugal in the Global Economy

Globalisation, that most favoured term of policy makers, commentators and academics at the beginning of the twenty-first century, has particular resonance within Portugal. The notion of globalisation is a profoundly contested one (Held et al, 1999). However, it is perhaps most commonly conceived of in terms of the deepening and extending of the networks of relationships that tie people, businesses, institutions and regions together in the same global space. From this standpoint globalisation attempts to grasp the economic, political and cultural transformation of modern society resulting from increased global interconnectedness, such that flows of goods, information, ideas, capital and people are more intense, occur more rapidly, and across a wider geographical space.

Many of the debates surrounding globalisation have focused on the claims from pro-globalisation commentators of the irreversibility of globalisation processes and the benefits of economic growth and political democracy resulting from a global shift towards a free trade, capitalist global economy and the establishment of western traditions of political democracy, freedom and citizenship. However sceptics raise a number of fundamental questions concerning this 'hyperglobalist' account of globalisation (for example Gray, 1998; Hirst and Thompson, 1996). Is globalisation really anything so different to the internationalisation processes which have evolved over the last 500 years and in which Portugal was such a primary actor in their earliest phase? Do not nation states still retain a key role in the management of national economic spaces and in seeking to control global economic forces? Is it not possible to construct alternatives to the inevitable triumph of a free market ideology and consumerist culture that exacerbates inequality, social polarisation and environmental degradation?

Such questions are highly pertinent to the case of Portugal. There can be no doubt that Portugal has become more outward looking and globally interconnected over the last thirty years, although the nature of this global integration has been strongly mediated through membership of the European Union, and heavily shaped by its history as a former colonial power. The benefits and difficulties of Portugal's emerging insertion into the global economy need therefore to be rigorously analysed and critically debated, perhaps in a manner that was sometimes absent during much of the 1980s and 1990s.

Corkill (Chapter Two) explores the changing nature of integration of the Portuguese economy into the European and global economy. He points to the key role played by Portugal in the historical processes leading to the creation of the global economy, but also the degree of isolation which Portugal experienced from international economic developments for a large part of the twentieth century. In tracing through Portugal's growing international integration, Corkill identifies how internationalisation processes led to major changes in the productive structure and the key role played by European integration. Europe has become the dominant focus for trade in terms of both imports and exports and was the major source of foreign direct investment (FDI) that helped stimulate economic growth in the 1980s and 1990s. However, as concern about the nature of Portugal's global economic integration grew in the 1990s, Portugal was keen to claim a new place for itself, building upon its former historical relationships. In particular, the idea of being a key link between Europe and Mercosur developed and accompanying its new found confidence, flows of Portuguese foreign direct investment increased and a number of major Portuguese multinationals emerged.

The changing nature of global integration is highly evident within the Portuguese tourist industry, an industry critical to economic development processes within Portugal and one deeply implicated within tendencies towards increased economic and cultural globalisation. In his analysis of the evolution of tourist development in Portugal, Williams (Chapter Four) stresses the importance of the nature of its international integration as well as its rootedness within particular localities. The Portuguese tourist industry continues to exhibit a strong dependence on particular source countries (UK, Germany, Spain), the predominance of mass summer tourism, and a spatial focus within a small number of centres (Algarve, Lisbon and Madeira). The increased international competition and downward pressure on prices faced by traditional forms of mass summer tourism has precipitated attempts to move the tourist experience up market, both to differentiate from other areas and to increase the tourist spend. Diversification into rural tourism, beach tourism on the Atlantic coast, short city breaks, and the rise of retirement migration and domestic tourism are therefore important new trends, although these market segments still account for only a relatively small amount of total tourist activity.

In his discussion of tourism, Williams identifies how tourism is linked into broader systems of international and domestic population mobility. In the case of Portugal this is exemplified by the importance of the role of return migrant flows from France, internal urban-rural flows, flows of business travel, and more recently, the increasing importance of retirement migration, particularly to the Algarve. Eaton (Chapter Five) examines the repositioning of Portugal within international population flows in further detail. His analysis of recent trends in international population mobility demonstrates how Portugal has moved from a country associated with large-scale emigration to one typified by a new phase of immigration. Although foreign residents and immigrant labour from Africa, South America and northern Europe remain a small component of the national population and workforce, they are increasingly significant, both because of their concentrations spatially and in particular labour markets, as well as via the new policy challenges they present in terms of immigration and race relations.

The impacts of economic globalisation within Portugal are differentiated sectorally, spatially and between different types of capital (Syrett, 1996). The sectoral dimension is well exemplified by Sousa's (Chapter Six) examination of the liberalisation of the media and communication sectors. Development in these sectors illustrates how Portugal's encounter with globalisation tendencies has often been heavily mediated by its membership of the European Union (Simões, 1993). Modernisation processes here were strongly driven forward by EU policies particularly with respect to the telecommunications sector; policies which were themselves attempts by the EU to strengthen its global position. However, Sousa also provides a nuanced account of the differences between the media and telecommunications sectors and the role of different forms of capital within them, and illustrates the crucial role of the Portuguese state in taking forward these changes in a manner which sought to liberalise activity yet retain state control.

The spatial impact of increased global economic integration is strongly interrelated to the dynamics of sectoral change. Syrett (Chapter Three) explores the interrelationships between sectoral and spatial change within the Portuguese context and identifies how the current phase of economic development has often reinforced traditional economic imbalances between the developed coastal belt and the interior, as well as creating new centres of growth and decline. Both Williams (Chapter Four) and Eaton (Chapter Five) demonstrate for tourism and immigration respectively, how processes of change are rooted within, and interact with, existing spatial distributions, and have tended to reinforce existing patterns of spatial development. Furthermore, as Pereira da Silva notes (Chapter Seven), it is the highly developed coastal region which is beset by the most serious environmental problems. Whilst for the asset rich metropolitan centres of the coastal belt globalisation processes may offer significant economic development opportunities, for less developed rural economies and traditional manufacturing areas, global integration is marked by increased threats of international competition in traditional economic sectors and the possibility of economic marginalisation.

As the contributions to this volume demonstrate, Portugal's interaction with the deepening and widening of global flows reveals the complex, contradictory and incomplete nature of the diverse set of processes that are untidily bundled together under the notion of globalisation. Economic globalisation is no simple, unilinear and coherent project and there is nothing inevitable about the eventual outcomes (Dicken et al, 1997). Rather globalisation tendencies are marked by discontinuities and unevenness of impact between regions, sectors, peoples, institutions and businesses. Analysis of the interplay of globalisation tendencies within Portugal must therefore recognise differences between places and peoples and remain sensitive to the particularities of history, politics and culture.

Although increased global economic integration has brought benefits to Portugal in terms of financial flows and access to new markets for expansionary indigenous businesses, the danger is that the overly narrow basis of its international integration (e.g. tourism, clothing and footwear, car components) makes the economy highly vulnerable to changes in external demand. The economic challenge of globalisation, as Corkill notes, is to relocate Portugal from a low cost producer to one that can compete on the basis of quality, design and innovation. Yet globalisation processes also serve to increase social polarisation, social exclusion and environmental degradation. As Pereira da Silva (Chapter Seven) demonstrates, environmental concerns have been consistently subordinated to the needs of an unsustainable economic growth process. Managing the social and environmental dimensions presents a further set of challenges; ones which have often received only limited attention in the vigorous pursuit of economic modernisation, liberalisation and global integration.

Reconfiguring the Role of the State

Strongly related to discussions of processes of economic globalisation have been debates over the changing nature and role of the nation state. Globalisation processes challenge the traditional powers and role of the nation state by reducing its role as the principal governance authority in two ways (Amin and Thrift, 1997). First, via the increased significance of supranational agencies and institutions at the global level, and of regional economies at the sub-national level (Jessop, 1994). Second, globalisation processes require nation states to reorientate their priorities towards securing global competitiveness. Traditional areas of existing nation state policy are weakened by supranational agencies and agreements (e.g. trade, monetary and industrial policy), whilst domestic economic policy increasingly turns away from traditional concerns with welfare to focus on issues of promoting national competitiveness within a global economy. Despite such changes the nation state retains responsibility for key elements of economic development and social welfare (e.g. education and training, provision of infrastructures, heath, social security etc.). The current period of economic and political transformation is therefore characterised by the reconfiguring of the nation state role rather than necessarily its diminishment.

There can be no doubt of the importance of supranational governance structures upon the development of the Portuguese nation state over the last twenty years. Accession to the EU alongside participation in international trade agreements has been critical in promoting the liberalisation of formerly protected markets and provided the context for the withdrawal of the state from substantial areas of economic activity, most notably via the extensive privatisation programme of the 1990s (Corkill, 1999; and see Sousa, Chapter Six). Participation in European Economic and Monetary Union has resulted in the relinquishing of control of key elements of economic and monetary policy to the European Central Bank and the disappearance of the *escudo* in

February 2002. The orientation of economic modernisation policies towards concerns with international competitiveness via improved levels of productivity, labour skills, and innovation was evident across successive governments of the Social Democrat Party (PSD) and Socialist Party (PS) in the 1980s and 1990s, and such concerns dominate the Third Community Support Framework which operates until 2006. Discussions in this volume on areas as diverse as regional policy, the liberalisation of media and communications, and environmental policy (Chapters Three, Six and Seven), all illustrate the key role EU policies, regulations and funding have played in shaping the respective policy agendas.

Yet whilst membership of the EU in particular has provided considerable pressure for the rapid reshaping of state roles and institutions, the legacy of past state practices rooted in the *Estado Novo* remains strong. The historical development of the Portuguese State provides strong traditions of centralisation and bureaucracy and weak traditions of democratic government, effective policy making and public administration, and welfare state functions. There is therefore a considerable tension present between pressures to adopt and develop state roles in line with other advanced European nations, but without a basis in the well-established state traditions of democracy, welfare and decentralisation present within the majority of these countries.

As Robinson (Chapter Eight) demonstrates, the establishment of a stable democracy given the lack of a democratic tradition and the political turmoil of the immediate post-revolutionary period, was a major achievement. The evolution of the constitutional system reflects the political and social evolution of Portugal towards its EU partners with the 1989 constitutional changes removing the commitment to socialism and completing the transition to a pluralistic, participative democracy. The political stability which emerged via the evolution of the major political parties and the continuity of leadership provided by two Presidents (Soares and Sampaio) and two Prime Ministers (Cavaco Silva and Guterres) from the mid 1980s, was critical to underwriting the dramatic socio-economic modernisation which characterised this period. Furthermore, the propounding of a neo-liberal agenda in the 1980s by the Cavaco Silva administration, and of 'Third Way' politics in the 1990s by the Guterres administration, ensured that Portuguese national politics sat comfortably in accordance with the dominant international political ideologies of the last 20 years.

Yet despite the rapid embedding of constitutional democracy the persistence of traditions of clientelism, elitism, bureaucracy and centralism serve to undermine the democratic process and the quality of political decision making (Magone, 1997; Corkill, 1997). These elements are clearly evident in Pereira da Silva's examination (Chapter Seven) of the development of environmental policy, perhaps most strikingly in the debacle over the Vasco de Gama bridge. As the case of environmental policy illustrates, in many cases it is not the absence of relevant policies which are the problem, rather their ineffective delivery and enforcement. Sousa's analysis of the reform of the media and communication sectors similarly reveals the clientelism, lack of transparency and strong degree of control by the political elite, which characterised the reform process and provides interesting evidence of continuity with past state practice. The continued distance between the political elite and the electorate has been a major factor in the falling electoral turn outs identified and discussed by Robinson (Chapter Eight), Silva (Chapter Nine) and Corkill (Chapter Ten). It has also sponsored the growth of alternative non-party political social movements such as the rapid rise in environmental groups (see Chapter Seven), oriented towards improved quality of life issues.

Whilst the Portuguese state remains highly centralised by European standards the establishment of elected local government in the post-revolutionary period marks a profound change. Silva (Chapter Nine) charts the recent evolution of the competencies, functions, financing and means of operation of local government. These changes have resulted in the establishment of democratically elected local authorities, which have a significant level of autonomy and operate across a wide range of policy areas. In particular Silva notes the emergence of new forms of governance at the local level, with the rise of public/private partnership and more innovative and entrepreneurial forms of organisation in order to improve the delivery of local services. With the rejection of the proposals for the creation of regional administrations in 1998, the pressure for greater autonomy, subsidiarity and flexibility at the municipal level increased further. Yet the ability of local government to pursue these goals remains fundamentally constrained by central government, most importantly because of the continued low level of total public administration expenditure (less than 10 per cent) accounted for by local government, a figure well below the European average. The ongoing strengthening and reform of local government appears to be a basic requirement if the Portuguese state is to improve the quality and efficiency of its delivery of a range of public services, as well as to reinforce notions of citizenship and local level participation within the political process (Ruivo, 2000).

The underdeveloped nature of welfare state functions provides further evidence of prevailing tensions in the evolution of the Portuguese state. From the late 1960s, Portugal witnessed a rapid growth of the welfare state with universal availability of services in education, health and social security (Carreira, 1996). Public spending as a percentage of GDP grew from 20.0 per cent in 1960 to 53.0 per cent in 1990. However the quality and level of public services remains weak reflecting the longstanding lack of capital investment, the weak administrative systems, and the low levels of skill and productivity of many public sector workers. Despite the commitment of successive government administrations to public sector reform, overall the number of public sector employees continued to rise across the 1990s to over 700,000 by 2000, absorbing the equivalent of 15 per cent of GDP, far higher than the European average.

The financing of welfare services within Portugal exhibits particular problems relating to its historical evolution (Barreto, 1996a). First, the majority of those who are dependent on the current social security system have not made financial contributions across their lives, a point particularly important given the rapidly ageing population. Second, the low productivity of enterprise means that business generates limited wealth for social investment. Third, relatively low levels of personal wealth means that there is limited capacity for individuals to pursue personal solutions via savings and insurance, in order to look after their own personal welfare. Finally, despite these limitations, expectations from the population are increasingly in line with those from more wealthy and developed societies, leading to a consequent gap between expectations and the economic and social capacities available to deliver them. These difficulties pose profound questions as to how the post-revolutionary commitment to the universal provision of public services and equality of social rights will be paid for in the future. Given that a period of strong economic growth has failed to generate enough income to fund the huge growth of state welfare functions, any future downturn in economic growth rates is set to exacerbate this situation.[5]

Conclusions

The difficulties of taking forward the development of the Portuguese welfare state illustrate more general fears concerning the sustainability of recent economic growth and the appropriateness of state institutional structures and capabilities. Given the failure of many state reforms in the 1980s and 1990s, concerns persist about the ability of the refashioned and enlarged state to function effectively given a rapidly changing external environment, rising domestic expectations, and profound social problems related to rising inequalities, poverty, social exclusion and environmental degradation. With the initial phase of rapid convergence to European norms largely successfully completed, the question for the future is whether it is possible to sustain such a rapid pace of change in the early decades of the twenty first century, particularly under less favourable economic circumstances. It seems unlikely that Portugal will be able to ignore the inherent tensions and paradoxes of rapid modernisation and global integration in the future, as it has done in the recent past. Confronting these issues will profoundly test the strength of the foundations on which contemporary Portuguese economic, social and political life has been built.

Notes

1   Recent years have seen a major increase in problems of drug abuse and associated criminal activity. By 2001 the Portuguese population had one of the highest rates of drug addiction in Europe, a factor which led the government to seek new approaches to tackling the problem via the decriminalisation of soft and hard drugs for personal use. Rising crime levels are strongly related to problems of drug abuse, with 70 per cent of inmates having committed drug-related offences.

2   For example the long-term decline in fishing-related employment intensified further between 1990-1996, falling from 40,500 to 28,500. As a result by the late 1990s this sector accounted for around only one per cent of total employment and two per cent of total national output by value.

3   See Eaton and Pereira da Silva (1997) for a discussion of the persistent problem of child workers in Portugal.

4   For example see Porter's (1990) highly influential book The Competitive Advantage of Nations. Michael Porter's consultancy company, Monitor, undertook a major analysis of the competitive advantage of the Portuguese economy in the early 1990s (Monitor Company, 1994) which subsequently informed the development of a 'cluster based' national economic strategy.

5   The international economic downturn in 2001 led to substantial public spending cuts of 750 million Euro, which included the suspension of young men being called up for compulsory military service. The Portuguese government was forced to act, albeit belatedly, when the budget deficit rose to 1.7 per cent of GDP in 2001 as they are committed to maintain a budget deficit at 0.7 per cent of GDP as part of the Eurozone stability pact.

References

Amin, A. and Thrift, N. (1997), 'Globalization, Socio-Economics, Territoriality', in R. Lee and J. Wills (eds) *Geographies of Economies*, Arnold, London, pp.147-157.

Baklanoff, E.N. (1978), *The Economic Transformation of Spain and Portugal*, Praeger, New York.

Barreto, A. (1996a), 'Três Décadas de Mudança Social', in A. Barreto (ed) *A Situação Social em Portugal, 1960-1995*, Instituto de Ciências Sociais, Universidade de Lisboa, Lisbon, pp.35-60.

Barreto, A. (ed) (1996b), *A Situação Social em Portugal, 1960-1995*, Instituto de Ciências Sociais, Universidade de Lisboa, Lisbon.

Carreira, H. M. (1996), 'As Políticas Sociais em Portugal', in A. Barreto (ed) *A Situação Social em Portugal, 1960-1995*, Instituto de Ciências Sociais, Universidade de Lisboa, Lisbon, pp.365-498.

Corkill, D. (1997), 'Interpreting Cavaquismo: the Cavaco Silva Decade in Portugal', *International Journal of Iberian Studies*, vol.10, no.2, pp.80-88.

Corkill, D. (1999), *The Development of the Portuguese Economy: a Case of Europeanization*, Routledge, London.

Dicken, P., Peck, J. and Tickell, A. (1997), 'Unpacking the Global', in R. Lee and J. Wills (eds) *Geographies of Economies*, Arnold, London, pp.158-166.

Eaton, M. and Pereira da Silva, C. (1997), 'Portugal's Child Workers', in *International Journal of Iberian Studies*, vol.10, no.3, pp.160-169.

Financial Times (2001), 'Euro Fuels Big Consumer Spree', *Financial Times Survey*, Annual Country Report, October 24, p.II.

Gallagher, T. (1983), *Portugal: A Twentieth Century Interpretation*, Manchester University Press, Manchester.

Graham, L.S. and Makler, H.M (eds) (1979), *Contemporary Portugal: the Revolution and its Antecedents*, University of Texas Press, Austin and London.

Graham, L.S. and Wheeler, D.L (eds) (1983), *In Search of Modern Portugal*, University of Wisconsin Press, Wisconsin.

Gray, J. (1998), *False Dawn: the Delusions of Global Capitalism*, Granta, London.

Held, D. and McGrew, A., Goldblatt, D. and Perraton, J. (1999), *Global Transformations*, Polity Press, Cambridge.

Hirst, P. and Thompson, G. (1996), *Globalization in Question: the International Economy and the Possibilities of Governance*, Polity Press, Cambridge.

Hyland, P. (1997), *Backwards Out of the Big World*, Flamingo, London.

Jessop, B. (1994), 'Post-Fordism and the State', in A. Amin (ed) *Post-Fordism: A Reader*, Blackwell, Oxford, pp.251-279.

Magone, J. (1997), *European Portugal. The Difficult Road to Sustainable Democracy*, Macmillan, Basingstoke.

Minstério do Planeamento (2000), *Quadro Comunitário de Apoio III: Portugal 2000-2006*, Ministério do Planeamento, Lisbon.

Monitor Company (1994), *Construir as Vantagens Competitivas de Portugal*, Forum para a Competitividade, INETI, Lisbon.

Porter, M.E. (1990), *The Competitive Advantage of Nations*, Macmillan, London.

Robinson, R.A.H. (1979), *Contemporary Portugal: A History*, Allen and Unwin, London.

Ruivo, F. (2000), *Poder Local e Exclusão Social*, Quarteto, Coimbra.

Santos, B de Sousa (ed) (1993), *Portugal: Um Retrato Singular*, Afrontamento, Porto.

Simões, V.C. (1993), 'Going Global or Going European? The Case of Portugal', in M. Humbert (ed) *The Impact of Globalisation on Europe's Firms and Industries*, Pinter, London, pp.152-172.

Sousa Ferreira, E, and Opello, W.C (eds) (1985), *Conflict and Change in Portugal 1974-84*, Teorema, Lisbon.

Syrett, S. (1996), 'Internationalisation, European Integration and the Portuguese Economy: the Case of the Food Manufacturing Sector', in *The Cyprus Journal of Economics*, vol.9, no.1, pp.66-88.

2 Portugal's Changing Integration into the European and Global Economy

DAVID CORKILL

Introduction

Over the past half century Portugal has moved along a development path that has taken it from the periphery towards the centre of the European core. The trajectory has not always been linear and at times there have been diversions and phases when progress stalled, not least because for many years Portugal tried to forge a separate identity based on its possession of a world-wide empire, an Atlanticist vocation, and an insular vision encapsulated by a preference for standing, in Salazar's famous phrase, 'proudly alone'.

It would be wrong to simply dismiss pre-1960s Portugal as 'peripheral', 'underdeveloped' and 'dependent'. In the twilight years of the dictatorship a more accurate categorisation would be to locate it among the newly-industrialising economies (NIEs) which, through export promotion and encouragement offered to foreign investors, began to modernise and become integrated into the international economy. Moreover, it is misleading to talk about Portugal as a latecomer to internationalisation. Centuries of isolation, backwardness and marginal status served to obscure the fact that the country was once a world power that had helped to shape the nascent global economy during the fifteenth and sixteenth centuries. Portugal can lay claim to be the first European country to build an Atlantic empire, based on Brazil, which was gradually extended to encompass the Indian Ocean and Far East. The Portuguese introduced the plantation system to the New World, and became one of the earliest global trading nations. It is, perhaps, more accurate to define Portugal's twentieth century marginality in terms of isolation from mainland Europe and some, but by no means all, of the major international institutions established in the wake of the Second World War.

Insertion into the International Economy

In trading terms Portugal remained peripheral to the European mainstream, exporting a limited number of products and reliant on a few trading partners, particularly Great Britain. Integration into the modern international economy did not begin in earnest until the 1960s when the politics of autarky and economic nationalism pursued by the Salazar regime were gradually abandoned. It should be emphasised that Portugal's 'opening up' to the international economy was a piecemeal process. Protectionist features remained in place, foreign investment was not welcomed unequivocally, and the state continued to play a directive role in the economy. A number of stages can be identified in the insertion process, corresponding to 1959-early 1970s, 1972-85, 1986-98 and post-1999 periods (see Table 2.1).

Phase One: Post-Autarkic Insertion

In 1959 Portugal joined EFTA (European Free Trade Association) and, as a poorer member, was allowed to retain protection for some of its industries and given an extended timetable for tariff reductions. Within a few years Portugal had become a member of GATT (General Agreement on Tariffs and Trade), the IMF (International Monetary Fund), and the World Bank.

These changes reflected the influence of the technocrats who were in charge of economic policy. They were more European-oriented and envisaged foreign investment playing a greater role in their growth strategy (Baklanoff, 1992, p.5). A series of development plans aimed to accelerate economic growth, at least in part because the regime needed to fund the

Table 2.1 The Insertion Phases, 1959-99

First phase	1959-early 1970s	EFTA, World Bank, IMF(1960), GATT (1962)
Second phase	1972-85	Trade agreement with EC9 (1972)
Third phase	1986-99	EC accession (1986), ERM (1992) Iberian market
Fourth phase	post-1999	Eurozone member, business internationalisation

Source: Author's periodisation

burgeoning military expenditure on the African wars. Salazar even sought external credits and loans from US, German and French banks during the 1960s, something he had studiously avoided in the past.

While pursuing closer relations with western Europe in the 1960s, the Salazar dictatorship also promoted a closer economic and administrative integration with the colonies. This involved population transfers, a common currency designated as the 'escudo area', trade liberalisation, and the encouragement of private investment flows from both Portuguese and foreign sources (Baklanoff, 1992, p.6). The African commitment and the protracted colonial wars (1961-74), fought to preserve the Portuguese presence on the continent, diverted valuable resources and directly contradicted the longer term economic logic. During the 1960s and early 1970s Portugal was stranded at a halfway house with 'its head in the First world and its feet in the Third', being both European and peripheral at the same time. This situation prevailed as long as Salazar, along with his successor Marcello Caetano, clung to the belief that a 'Portuguese economic space', comprising Portugal's 'overseas provinces' in Africa and Asia, could be a viable alternative in the country's future political economy.

There was some logic to the dictatorship's determination to retain its empire well into the post-imperial era. The colonies continued to be important suppliers of raw materials and provided a market for Portugal's exports. Indeed, they were vital for those manufacturers producing inferior quality, uncompetitive goods. But, while there was trade growth in the escudo area,

Table 2.2 Distribution of Portuguese Exports and Imports (%)

	1973		1985		1995	
	Exports	Imports	Exports	Imports	Exports	Imports
European Union	48.6 (EC9)	44.9	58.1 (EC12)	48.8	80.1	73.9
NAFTA			10.0	12.2	5.3	4.0
EFTA	13.8	11.6	10.8	6.0	3.1	2.9
PALOPs	14.8	10.1	3.9	1.2	2.5	0.2
Asia	–	–	1.6	4.3	2.0	5.0
Mercosur	–	–	0.9	3.6	1.1	2.0
OPEC	–	–	2.5	17.1	0.7	5.1

Source: INE; Silva Lopes, 1996, p.164

the *ultramar's* share in total trade contracted during the 1960s and early 1970s. By contrast, Europe's role as a supplier and as a market increased exponentially (see Table 2.2). This growing dependence on western Europe was increasingly apparent in a number of other areas. First, Portugal became a major supplier of labour to the booming economies in northern Europe. Between 1964-74 some one and a half million migrants quit the country to find work abroad; giving rise to the observation that Portugal's 'greatest export was its people'. Second, as a direct consequence of this labour migration, emigrant remittances (ERs) became a major element in the balance of payments, facilitating economic development initiatives and enabling Salazar to fight the colonial wars without incurring large scale indebtedness. On average ERs totalled some $1.1 billion by the early 1970s, representing about 10 per cent of national income. There were two other important integrative mechanisms. Tourism began to generate substantial foreign currency earnings, and foreign direct investment, attracted by the cheap labour, transformed the economic structure and boosted exports.

Phase Two: Insertion into the European Space and Industrialisation

The 1972 trade agreement reduced the barriers between Portugal and the EEC (European Economic Community). It represented an acknowledgement that the EEC's second enlargement (when Britain, Ireland and Denmark joined) could have negative consequences for the country's trade. Portugal now had two major economic arrangements with western Europe which, significantly, extended the markets for Portuguese goods and provided a further stimulus to the diversification and modernisation of Portugal's economic fabric.

Closer integration with EFTA and the EEC triggered some major developments in Portugal's economy. First it marked the start of the gradual dismantling of barriers to trade between Portugal and Europe. This process consolidated the position of Europe as the chief market for Portugal's exports and the major source of its imports. Second, the foreign trade structure diversified away from the traditional reliance on a limited range of agricultural products to include a greater share of manufactured exports (Baklanoff, 1992, p.11). From a small and restricted product range (canned fish, cork, textiles, wine) Portugal began to export machinery and transport equipment, including motor cars, as well as pulp and paper, clothing and footwear, and chemicals. New export-oriented industries emerged in the late 1960s/early 1970s as Portugal joined the ranks of the newly industrialising economies.

Phase Three: EC Accession

Portugal applied to become a member of the European Community in 1977 and eventually joined in 1986. EC accession accelerated the established trend towards closer economic relations with Europe. Portugal signed the Maastricht Treaty in 1992 and implemented the convergence criteria that were prerequisites for participating in the euro project launched in 1999. By the late 1990s trade concentration had developed to the point where 80 per cent of Portugal's foreign trade was with EU countries. In the period between 1980 and 1997 exports to the EU rose by 15 per cent, while imports surged by 27 per cent.

An influx of European Structural Fund monies in the form of pre- and post-accession aid boosted infrastructure investment and underpinned rising business confidence. A contributory factor was the stimulus provided by the introduction of the Single European Market (SEM) in the early 1990s. However, closer integration did expose the essential dualism and vulner-ability of Portugal's economic structure. Traditional sectors reliant on cheap labour and low technology continued to co-exist alongside more advanced, high-tech ones. Integration brought increased opportunities for improving market share but, by the same token, exposed the uncompetitiveness and vulnerability of many sectors and businesses.

EC accession also marked a significant increase in Iberian trade relations. The EU's southern enlargement laid the basis for the emergence of an 'Iberian market'; a regional subunit of the larger EU trading block. By the late 1990s Spain had taken over as Portugal's principal supplier, and the Iberian neighbour was the second most important export market for Portuguese goods. As a consequence, the relative positions enjoyed by EFTA, the USA and other non-European countries declined markedly.

This period was also characterised by the growing importance of flows of foreign direct investment (FDI) into the Portuguese economy. Encouraged by a new investment code (1986) which relaxed entry require-ments and introduced favourable changes in the contract regime, foreign capital moved into the manufacturing, financial services and tourism sectors. In so doing, foreign investors aimed to take advantage of access to the large European Union and growing domestic markets, relatively low labour costs, generous financial and tax incentives, and improving infrastructures. The government's large scale privatisation programme set in motion in the late 1980s (Corkill, 1999) provided additional investment opportunities for foreign investors. As a result, FDI, which amounted to 0.6% of Portugal's

GDP in 1986, had risen to 8.1 per cent by 1993. The opening of financial markets after 1993 further encouraged foreign investment projects which aided restructuring and modernisation efforts.

The development of export-oriented FDI projects gave additional impetus to the penetration of export markets. The largest and most high profile export-platform type investment was the AutoEuropa plant, built at Palmela in the 1990s. Initially a Ford-VW joint venture (now fully owned by VW) the plant produced the Galaxy/Sharan multipurpose vehicle. Vehicle exports surged and Germany became the chief market destination for Portuguese exports (27 per cent of total exports in 1995). As a result, high unit value products increased their share of total exports, thus further underlining the nature of the transformation. By the late 1990s, the company accounted for 11 per cent of Portugal's exports and 2.1 per cent of GDP.

Portugal's European vocation had been confirmed following the collapse of the empire in 1975 which left Portugal with little choice but to accept a subordinate role and integration into the Community. The only possible alternative was Lusophone Africa, but Angola, Mozambique and Guinea-Bissau were torn apart by decades-long civil war and instability. However, recent improvements (partly undermined by renewed fighting in Angola) have bolstered the case of those who argue that Africa is still important for Portugal. Some efforts have been made to lay the foundations

Table 2.3 The Community of Portuguese-speaking Countries (CPLP)

	Population (millions)	GDP per capita (US$, 1996)	Economic links
Angola	11.5a	772	SADC
Brazil	159.1	4,772	Mercosur
Cape Verde	0.4a	1,214	Francophone
Guinea-Bissau	1.1	189a	Francophone
Mozambique	16.1b	129	SADC
Portugal	9.9	10,417	EU
St Thomas Islands	0.1a	252	Francophone

All figures are for 1994 except for a1995 and b1991

Source: Euromonitor, 1997

for closer Lusophone economic relations, although the Community of Portuguese-speaking Countries (CPLP) remains fundamentally a cultural rather than an economic organisation (see Table 2.3).

Phase Four: Joining Euroland

The prime minister, António Guterres, long maintained that European Monetary Union (EMU) was an instrument to ensure that Portugal remained at the heart of the European project. Consequently, Portugal became committed to being among the 'first wave' members of the new single currency launched in 1999. Membership of a large economic space was perceived to be the best chance to overcome backwardness and converge with the more prosperous partners. Above all, participating in the process of economic and monetary union carried symbolic significance for Lisbon because it signalled that the country had joined the European mainstream and was no longer a marginal state and latecomer moderniser. Despite the success in meeting the criteria for monetary union and joining the 11 countries at the euro's launch, this latest insertion phase was not without its dangers for Portugal. Locking a disparate group of countries together in a eurozone took Europeans into uncharted territory. The danger existed that the peripheral countries in the euro area (Portugal, Spain. Ireland and Finland), who were growing faster than the core economies during the 1990s, might overheat. In one scenario problems could be stoked by excessive borrowing and indebtedness, soaring asset values and excessive wage demands.

Whilst greater European and international integration has brought many benefits to the Portuguese economy there are also worries that Portugal has become overreliant on external stimuli. For instance, it can no longer be taken for granted that transfers (emigrant remittances and EU funds) will compensate for the growing current account deficit. Although it is a sign of economic maturity that investment outflows are causing concern and ERs declining, it does prompt disquiet over the deficit in the trade and invisible account. The fear is that the Portuguese economy has become overdependent on output from tourism earnings and the AutoEuropa plant, both of which are vulnerable to the vagaries of external demand and ever-fiercer market competition. The AutoEuropa case underlines the dangers. Ford's decision to pull out of AutoEuropa because of overcapacity in the motor industry means that production of the Ford Galaxy will end in 2004. A replacement model has yet to be found. Hopes had been pinned on the new VW-Porsche Sport Utility Vehicle, but the decision went in favour of

Leipzig in Germany. Even more sobering is the knowledge that in the VW Group plants producing less than 100,000 units or more are regarded as uncompetitive. Production volumes for the Sharan (VW) and Alhambra (Seat) totalled under 78,000 in 1998, less than half their original capacity (Diário de Notícias, 1999).

With regard to EU Structural Funds, the Third Community Support Framework (2000-2006) will almost certainly be the last occasion that Portugal benefits on such a scale. The next enlargement phase is likely to disadvantage Portugal in a number of ways. First, the influx of new members means that, as the budget allocated to the Structural Funds is unlikely to be increased, there will be far less to go around. Second, the new entrants will lower the EU's average GDP, placing Portugal closer to the Community average. Finally, the countries about to join not only compete directly with Portugal in terms of their exports, but also offer lower wages and well-trained workforces.

Internationalisation

EU integration accentuated the extent to which the economy was exposed to external competition. It meant that exchange rate manipulation, the traditional method used to promote exports prior to 1986, was no longer an option. In addition, cheap labour and other factors of production did not always suffice to guarantee access to foreign investment and world markets. Inevitably, the focus switched to improving competitiveness and productivity.

Since the beginning of the 1990s 'internationalisation fever' has gripped some sectors of the Portuguese economy. The press presented the process as a new 'age of discoveries', launched 500 years after Cabral landed in Brazil. In this case the new pioneers in the 'age of global-isation' were the Portuguese businessmen seeking investment opportunities in trade blocks such as Mercosur, the fourth most important regional trading area in the world. Optimistically some analysts envisaged Portugal becoming 'Europe's Atlantic platform' arguing, as did one leading Portuguese businessman, that South Americans prefer European culture to the 'hamburger culture' associated with the United States (Público, 1999a). Certainly the development of closer links with Latin American trading blocks like Mercosur has potential for the future. However, the risks were underlined during 1999 when a financial crisis hit the *real*, the Brazilian currency.

Inevitably Portuguese investments must take the long-term view.

There are certain features which are pertinent when assessing the prospects for successful internationalisation in Portugal's case:

- The economy has always looked outward and been more open than its neighbour, Spain. Nevertheless, there was a belated start to the internationalisation process. In a few small niches experience of internationalisation already existed; the best examples are port wine, table wine (Mateus Rosé, vinho verde), and cork. In contrast to its Iberian neighbour, Portugal's limited internal market obliged firms to look abroad and be more externally-oriented. There is an established tradition that small and medium-sized enterprises (SMEs) are export-focused. SMEs account for around half the country's total exports. In addition, many of the old groups and entrepreneurs from the Salazar era have returned, some from exile. They brought with them years of experience operating in competitive markets and important contacts with foreign investors and banks;
- An imperial past, 500 years of Portuguese colonialism, commerce and emigration have left an imprint around the globe. Little attention had been paid to the Portuguese communities until the 1980s when belated recognition was accorded to the part they might play in promoting economic development and assisting internationalisation. Large emigrant communities in the USA, South Africa, Venezuela, Europe (especially France) and the Portuguese-speaking world provided a focus for efforts in this regard (see Table 2.4). But inevitably, taking advantage of such a resource depends on being able to exploit this cultural reference (which, for instance, the Irish have been conspicuously successful in doing), and the comparative advantage that exists in terms of language, business culture and market knowledge;
- Historically, foreign investment played a major part in transforming the economic structure. Multinationals such as Texas Instruments, Inlan, Ford, Siemens, Pioneer, Blaupunkt established themselves in Portugal during the 1960s, but the high water mark came in the late 1980s with the largest ever foreign direct investment in Portugal in a leading sector, motor vehicles. The Ford-VW project proved to be a major exporter and job creator. Apart from this flagship project, foreign investors have displayed sustained interest in financial services and tourism. Generally-speaking, the two most common motives cited for decisions to invest were low production costs and

the generous incentives on offer. However, the record levels of FDI were not sustained in the late 1990s and inhibitors to foreign investment still existed. They included a poorly qualified workforce, infrastructure deficiencies, and limited research and development (R&D). All are factors which can adversely affect decision-making in the increasingly fierce international competition to secure inward investment;

- It has to be recognised that internationalisation efforts are in the relatively early stages. Due to the late start and the intensely competitive environment, the state has necessarily been involved in encouraging the process in selected sectors. ICEP (*Investimentos, Comércio e Turismo de Portugal*), the agency with responsibilities for internationalisation and export promotion, provided capital finance and credits. It has also co-ordinated efforts to improve the global reputation associated with Portuguese goods. The task has been to upgrade the 'made in Portugal' brand and affirm differentiation, replacing the cheap and low quality perception with an image based on quality and innovation. ICEP has encouraged co-operation among national firms to create critical mass in order to compete more effectively at an international level;

- The most recent development has been the increase in Portuguese investment abroad. During the 1990s opportunities multiplied at an impressive rate and by 1998 Portugal entered the ranks of the leading fifteen countries that invest overseas. The value of Portuguese investment overseas overtook the value of inward foreign investment; a measure of the progress that has been made since the 1980s (Wise, 1999, p.13).

Table 2.4 Portuguese Communities Overseas, 1992

The Americas		Europe		Africa, Asia & Pacific	
Canada	523,000	France	767,304	South Africa	600,000
USA	379,341	Germany	101,625	Angola	18,000
Venezuela	400,000	Spain	70,000	Mozambique	10,866
Argentina	18,000	UK	52,000	Macau	140,000
Brazil	1,200,000			Australia	60,000

Source: O Emigrante, 1994, pp.24-25

The Trajectory of Portuguese Foreign Investment

Between 1985-92 Portuguese companies focused primarily on the national market where there was strong growth in demand and consumption. Two factors caused them to re-orientate their strategies. First, the economic slow-down during 1992-93 compelled businesses to contemplate their first efforts at internationalisation. Second, some firms had simply hit the glass ceiling on further growth in the domestic economy and any expansion inevitably meant selling and investing overseas.

Internationalisation processes have usually followed a distinctive geographical pattern. Typically, the first moves were made in neighbouring Spain. Cross-border investments characteristic of this early stage were often unambitious, venturing no further than Galicia which enjoys strong cultural and linguistic affinities with Portugal. A signpost to the launch of this internationalisation phase came when Cimpor, the largest cement maker, acquired Corporación Noroeste, a Spanish cement maker based in Galicia. The next stage was for Portuguese firms as varied as Cenoura, Petrogal (rival to Elf and BP in Spain), Pão de Azúçar, Transportes Luis Simões and CGD to penetrate the Spanish market. Founded on rapidly expanding two-way flows, an EU regional 'Iberian market' emerged during the 1990s which provided opportunities for Portuguese investors but, conversely, also raised the spectre of Spanish domination of this buoyant new trading area. Using Spain as a springboard, the next natural extension for growth-minded Portuguese businesses was into North Africa and the countries that were formerly Portugal's African colonies (PALOPs). The latter became particularly attractive as they had launched privatisation programmes aimed at reconstructing their war-battered economies. Additionally, the opening of Brazil and the Mercosur market provided opportunities in the late 1990s. The most recent stage has been to invest in the more advanced EU economies and to take the first tentative steps into Eastern Europe, especially Poland, Hungary and Russia. Macao, Portugal's last remaining colony in Asia (until 1999), was regarded as a gateway for penetration into the Asian market, especially China.

This geographical pattern of economic internationalisation was confirmed in a survey conducted among Portuguese companies in 1997. It found that the priority markets for internationalisation were Spain (69.9 per cent), the PALOPs (47.3 per cent), rest of the EU (38.7 per cent) and Brazil (35.5 per cent) (Expresso, 1997). In recent years Portuguese companies have invested in excess of one billion (thousand million) contos in countries such as Brazil, Mozambique and Spain in an effort to gain a foothold in the global marketplace. Figures for 1998 indicated that some 80 Portuguese firms invested over 750 million contos abroad. Geographically the spread extended beyond the Portuguese-speaking world and Spain to include Poland, China, Morocco, and Tunisia. The European Union remained the main destination for Portuguese FDI (78.7 per cent of the total in 1990), but Brazil retained a significant share, absorbing more than $4 billion in 1998. The focus on Brazil as an investment outlet was underlined by figures indicating that the former colony absorbed 30 per cent of Portugal's total

---

Box 2.1   Sonae: Creating a Portuguese Multinational

Sonae is Portugal's largest company in terms of sales. It developed from a small family firm to become a major force in retailing and distribution led by Belmiro de Azevedo. The core business is dominated by the supermarket chains Modelo and Continente. However, as the demand for new supermarkets and shopping centres reached saturation point in the domestic market, Sonae has diversified into property, tourism, wood products, agribusiness and information technology (Corkill, 1999, p.140).

During the late 1980s Sonae mapped out its globalisation strategy when it realised that the growth possibilities in the domestic market had been exhausted. Competitiveness could only be achieved through establishing a presence in the global marketplace. Sonae's burgeoning wood products activities, given high transport costs but cheap raw materials, needed to locate as close as possible to the large markets. In pursuance of this goal Sonae acquired a factory in Northern Ireland and added further plants in Spain, Canada and England. During 1998 Sonae Indústria built new factories in Brazil, South Africa and northern England.

Sonae ensured its leading position in the world market for wood products by acquiring the second largest producer, Glunz, a loss-making German group with 25 factories based in Germany, France and England. Its retail operation (Sonae Retalho Especializado) took a similar expansionary route. In this case the focus was on the more prosperous states in Brazil. By 1998 Sonae had become the fifth largest retailer in the Brazilian market in direct competition with some of the supermarket giants like Carrefour (Santos, 1999, p.14).

foreign investment in 1996 and that the proportion has continued to climb since. In a major investment drive Portugal Telecom committed 450 million contos, and Grupo Cintra, which owns a mineral water company in Spain and five factories producing beers and bottled water in Brazil, has targeted Brazil for further growth.

Internationalisation strategies adopted by Portuguese business have been principally driven by the need to stay competitive in order to survive. As a result, Portugal began to establish a number of indigenous multinational companies (see Box 2.1). Expansions have taken the form of takeovers, joint-ventures or partnerships, funded from a company's own capital resources, medium and short term borrowing as well as by government-sponsored incentives for internationalisation. Given the constraints posed by the size of the domestic market, a key strategy for Portuguese firms to compete internationally was to merge with their indigenous rivals and form strategic alliances. The Portuguese banking sector was the first to experience this process. However this sector also illustrates the tensions which can arise out of the pursuit of economic strategies aimed at increasing size and market presence via consolidations and mergers with other national and foreign banks, and public policies driven by non-economic criteria. For example, in 1999 the Portuguese government vetoed the proposed merger between the Champalimaud group (banking and financial services) and the large Spanish bank Banco Santander Central Hispano (BSCH). Prime Minister António Guterres was determined not to allow the largest private financial group to fall into foreign, and particularly Spanish, hands. For a small country the nationality of corporate ownership can be an emotive issue, and hence also a politically sensitive one. The proposed cross-border merger sparked a confrontation between Lisbon and the European Commission's competition minister. It was only resolved when a complicated face-saving formula was devised. This example indicates that cultural and economic regions do not always coincide, and that there may not always be harmony at the level of the firm and the government in the quest for a more internationally competitive economy (Corkill, 2000, p.184).

Europe's Changing Economic Geography

At the European level the Portuguese economy faced challenges resulting from its location and specialisation. David Owen, an economist at Dresdner Kleinwort Benson, argued that based on the US experience, the boost to trade links generated by the EMU will inevitably lead to changes in Europe's economic geography. Owen predicted that there will be greater industrial specialisation across Europe with individual countries concentrating on fewer products. Using trade flow data he identified the winners and losers for each EU member state. In Portugal's case the best positioned industries were deemed to be alcoholic beverages, paper and packaging, household goods and mining. He concluded that most other industries suffered from a comparative disadvantage with general engineering, chemicals and pharmaceuticals emerging as the chief losers. According to this study, tourism and leisure, along with clothing and footwear, are likely to retain their importance in Portugal's economic profile well into the new millennium.

In the past Portugal's export clusters were based on natural resource or labour intensive activities. According to Owen, in future, advantages to be gained as a low wage location may be less than assumed by conventional economic wisdom. This is because investors are likely to focus their attention on sectors that have a comparative advantage, rather than simply seeking out locations that can offer cheap labour. In this scenario they will normally favour clusters that can offer an excellent infrastructure, specialist components suppliers and other factors that serve to offset wage savings (Owen, 1999, p.6-7).

Table 2.5 Competitive Disadvantages

State intervention:	excessive bureaucracy
	tax avoidance
Management:	low levels of entrepreneurial skills
	lack of enterprise culture
	weak management capabilities
Technology:	few links between business and universities
	low investment in R&D
Investment:	limited access to capital (especially SMEs)
	indebtedness/high interest rates
Education:	skills mismatch
	shortage of skilled human resources
Production:	low and inconsistent quality
	low tech methods
	inadequate investment

Source: Adapted from McDermott, 1997, p.495 and Bayão Horta, 1995, p.11

Above all, as a Dun & Bradstreet report underlined, human resources training and management strategies will be the key to success in the future (Expresso, 1999a). There is evidence that regrouping and consolidation has occurred in finance and banking, but the response has been slower in other sectors. This particular shift may come to pass in the medium to long term, but the evidence from Portugal suggests that investment decisions are still strongly influenced by factors such as low labour costs.

Competitive Disadvantages

Portugal has made substantial progress in implementing structural reforms that enhance competitiveness. Labour market flexibility ensured that the unemployment rate remained comparatively low. An extensive privatisation programme begun in the late 1980s reduced or eliminated state-ownership in sectors such as financial services, industry and public utilities. In addition, there has been a marked improvement in infrastructure endowment. These developments have taken place against a backdrop of political stability and continuity in macroeconomic policy.

However, a raft of problems remain to be tackled (see Table 2.5):

- to identify and promote competitive potential. It is important to diversify the export base in order to overcome the concentration on a few products and a limited number of markets. As a consequence of this exposure negative developments in the European and global economy impact markedly on the domestic economy. The dependence is particularly acute with regard to Germany which remains the leading market, followed by Spain, France and the UK. Trade with the PALOPs and South Africa is growing, but is constrained by non-economic factors.
- to resolve to what extent the protectionist impulse can be retained and 'national champions' nurtured by the state.
- to improve the quality of human resource skills and training. Portugal is vulnerable with regard to the educational levels attained by its workforce (the lowest in Europe) which hinders the ability to compete in some sectors.
- to remove red tape, excessive bureaucracy and poor linkages between the public administration and the private sector. In the business world the dominant corporate culture tends to impede change and innovation.

- Productivity requires urgent attention. Portugal has the lowest productivity levels in the EU, a little over 40 per cent of the European average, although this figure disguises substantial sectoral variation.

Identifying the Winners: a Competitiveness Audit

Changing integration into international markets has led to increasing discussion of Portugal's place in the global economy and the basis of Portuguese economic competitiveness (see Box 2.2). A competitiveness audit carried out in the early 1990s by Porter's Monitor consultancy recommended that Portugal should focus on a number of 'clusters' (Bayão Horta, 1995, p.76): motor vehicles, ceramics, cork, footwear, furniture, moulds, ornamental stones, textiles and clothing, wood products. Interestingly, many of the clusters selected were traditional industries with strong regional (primarily northern) roots. Almost all were in intensely competitive sectors at the European and global level. Apart from motor vehicles, all suffer from weaknesses with regard to marketing, sales and distribution. Human resources at both management and workforce levels are identified as poorly qualified and deficient in training. Furthermore, shortcomings existed in R&D, notably in the textiles, clothing and cork industries.

Porter recommended major changes in the production processes for the footwear, textiles, clothing and ornamental stone industries in order to make them more competitive. In fact, it is arguable that to offer a prescription to cover all sectors that require modernisation is inappropriate given the important differences between them. This sectoral variation is demonstrated in the following case studies which examine the international integration of three contrasting industries; motor vehicle components, wine and footwear.

Motor vehicle components   Prior to the opening of AutoEuropa, motor vehicle production focused mainly on assembly operations and many components suppliers simply did not match up to international standards. The Ford/VW project increased the number of certified suppliers. However, as many were small scale they lacked the capacity to expand their business abroad in what is a highly competitive market. Certainly AutoEuropa compelled suppliers to think in terms of 'lean' production, flexibility and standards. To this end 'suppliers clubs' were established (*Club de Fornecedores AutoEuropa*). Many components firms supplied the Palmela plant and exported the remainder of their output. By the mid-1990s the components industry employed over 23,000 workers, producing motors, transmissions, brakes, car bodies, suspension systems, interiors etc.

Motor components is a particularly interesting test-bed because manufacturers can turn to Eastern Europe and extra-Europe locations for sourcing. Evidently investors still appear to seek traditional advantages. Portuguese car components workers have the longest working week (42 hours) and wage costs that are 79 per cent below those in Germany, 68 per cent lower than in France and 55 per cent lower than in Spain. However, the Portuguese are caught in a pincer between gradually rising labour costs and new competitors who can undercut them by offering even lower wages.

Wine industry Porter confirmed that Portugal enjoys optimal climatic conditions and a skills base for competitive wine production. An accumulated export experience and knowledge of foreign markets exists, based on the centuries-old export of Madeira and port wines. Indeed, wine is the

---

Box 2.2 Portugal: Some Indicators

- According to the World Bank Portugal's GDP per capita stood at $14,380 in 1998, putting it in 45th place out of 210 countries. Among its European Union partners only Greece was placed lower down the rankings. When GDP is adjusted for purchasing power parities Portugal's economy slips down to 51st position. When indicators such as literacy and child labour are factored in, Portugal is on a par with countries like Sri Lanka and Algeria.
- Between 1990-98 GDP grew at an annual average rate of 2.3 per cent, lower than the 3.1 per cent recorded for the 1980s. Services showed the strongest growth, while industry grew by only 0.5 per cent and agriculture contracted by 0.4 per cent.
- Portugal has been steadily converging with its European partners since accession to the EC/EU in 1986. According to Eurostat GDP per capita stands at 74 per cent of the European average. The gap was reduced by 6 percentage points between 1991-95, but the 'catch up' rhythm slowed between 1995-99 registering a gain of around 3 per cent.
- According to the World Competitiveness Yearbook 1999 Portugal improved its position in the ranking by moving into 28th position, behind Malaysia and Hungary, but ahead of Italy and Greece.

Source: Information cited in Expresso, 1999b and 1999c

leading agricultural export, accounting for 3 per cent of total exports. Nevertheless, Portuguese table wines meet some resistance and require imaginative marketing because indigenous grape varieties are distinctive and less well known. Areas identified for attention include improving both Portugal's market image (more estate bottling and quality control) and upgrading the technical qualifications among wine producing personnel.

Port wine already has an international reputation, high standards and effective marketing. These positive attributes are not evident throughout the rest of the wine industry. Its potential has long been recognised, but there is too much 'lazy wine-making' pandering to the less demanding domestic market. Efforts are required to maximise the positive advantages enjoyed by the wine industry including the existence of seven indigenous grape varieties and excellent wine growing regions (Bairrada, Douro, Alentejo, Ribatejo etc.). The chief obstacle to the wines becoming more widely known is, once again, the 'country-of-origin' reputation which associates Portugal with Mateus Rosé (launched in 1942 and currently owned by Sogrape), and young table wines (vinho verde). This is exacerbated by the industry proclivity for identifying wines by type (Cabernet Sauvignon, Chardonnay etc.) rather than by country-of-origin. However some countries have successfully overcome this disadvantage, the most notable examples being Rioja (Spain) and Chianti (Italy). Clearly there is also much to be learned from 'new world' producers such as Australia, Chile and Argentina. Efforts need to be made in terms of product awareness in order to exploit the grape variety uniqueness (McDermott 1997, p.503). To this end Vini Portugal was established to improve the marketing of Portuguese wines and to promote internationalisation. Sogrape's decision to acquire Finca Fishman, an Argentine estate, in 1999 underlined that this company's ambitious growth and development plans are not limited to Europe.

Footwear The footwear industry is a particularly interesting case because it is a traditional industry noted for its poorly qualified workforce that has registered significant success in terms of enhanced competitiveness, job creation, growth in the number of firms and increased sales volume. Yet only a decade ago the industry was identified as a possible casualty of growing international competition. Adaptability to changing global market conditions has been a signature feature of the SMEs in this sector which explains its survival and the preservation of employment. Around 80 per cent of footwear production is destined for external markets and exports have grown at an average annual rate of 35 per cent, placing Portugal in the world top ten exporters and as the second most important in Europe.

Over the last 25 years employment has grown substantially from 15,000 to 50,000 workers. The industry is dominated by small family-owned businesses and micro-firms (83.4 per cent of the total). It is highly concentrated regionally with 90 per cent of production located in the northern districts of Aveiro, Oporto and Braga. According to 1994 data, 22 per cent of footwear workers had no qualifications and 46 per cent were only 'semi-qualified'. Almost 80 per cent had received only basic primary education. Surprisingly, the lack of educational success does not appear to have hindered competitiveness improvements. Despite an unpromising structure there are signs that the new technologies are being embraced and workers are becoming better qualified, demonstrating a remarkable capacity to 'learn on the job' (Público, 1999b).

Despite these positive signs, the pace of change is such that there is a continuing danger that productivity levels in the footwear industry will remain stubbornly below the chief competitors, Italy and Spain, and that rising labour costs will blunt its comparative advantages. The dumping strategies practised by Thailand, China and Indonesia pose a threat and have forced Portuguese producers to concentrate on quality upgrading and product differentiation. However, there have been dramatic improvements to the brand image associated with Portuguese shoes which previously suffered in comparison to Italy and Spain.

Conclusion: the Contradictions of Integration

Portugal is still some way from full convergence with its European partners. Per capita income stands at around 74 per cent of the EU average (Expresso, 1999b), while wages and salaries are still the lowest in the EU. This labour-cost advantage may enable Portugal to continue to outperform the larger Euroland economies for some years to come. The optimistic scenario is that Portuguese living standards will continue to rise as the gap between the richer core countries and the poorer fringe economies narrows over the next few years. A more sanguine view is that the pace of convergence is decelerating and it will be decades before Portugal is able to close the gap with its richer partners. Clearly key deficiencies remain in the areas of productivity, human resources and technology.

Given the country's dimensions, large-scale nationally-based groups are inevitably thin on the ground. World ranking tables, for instance, placed

Portugal's largest bank, the Caixa Geral de Depósitos, in 146th place. The danger always exists that Portuguese firms will be taken over by foreign, perhaps Spanish, transnationals. This poses a dilemma for public policy with regard to the industrial and service sectors. Given that size is an important factor in competitiveness, should the state support these groups, encourage mergers and expansion when the ultimate beneficiaries may be the shareholders and foreign groups? Or is it better to concentrate efforts and resources on small and medium-sized firms (SMEs)? Indeed, some economists argue that the priority should be to tackle the broad structural problems facing Portuguese business (deficiencies in human resources at both management and workforce levels for instance), rather than focus on a few sectors identified as candidates for competitiveness enhancement.

Although, as latecomer moderniser belatedly integrated into the EC/EU, Portugal only recently embarked on the process of modernising its institutions, policies and productive structures, considerable progress has been made to date. European integration played a key part in the rapid economic transformation that is under way and is expected to exert similar influence in the future. It is a measure of the advances made that Portugal now faces the task of compensating for losses in labour cost competitiveness with factors such as innovation, quality and design. The challenge is to continue the reconfiguration from an economy with a limited number of internationally competitive industries into a broader-based export-competitive one.

References

Baklanoff, E. (1992), 'The Political Economy of Portugal's later Estado Novo: a Critique of the Stagnation Thesis', *Luso-Brazilian Review*, vol.xxix, no.1, pp.1-17.

Bayão Horta, R. (1995), *A Competividade da Economia Portuguesa*, Forum para a Competividade, Lisbon.

Corkill, D. (1999), *The Development of the Portuguese Economy: a Case of Europeanization*, Routledge, London.

Corkill, D. (2000), 'Cross-Border Banking Mergers: the Case of Spain's BSCH and Portugal's Champalimaud Group', *International Journal of Iberian Studies*, vol.12, no.3, pp.173-84.

Diário de Notícias (1999), 'AutoEuropa luta pela sobrevivência', 27 September.

Emigrante (1994), 'Os Portugueses no mundo', 11 June, pp.24-25.

Expresso (1997), 'Espanha é a preferida', 18 November.

Expresso (1999a), 'A europeização não é benéfica', 6 March.

Expresso (1999b), 'Convergência com UE diminuiu', 8 February.

Expresso (1999c), 'Portugal mais competitivo', 11 June.

McDermott (1997), 'Competing from Southern Europe: the Case of Portugal', in B. Fynes and S. Ennis (eds) *Competing From the Periphery: Core Issues in International Business*, Oak Tree Press, Dublin, pp.475-520.

Owen, D. (1999), 'Economic Geography Rewritten', *Economic Focus*, January, Dresdner Kleinwort Benson, London.

Público (1999a), 'Prioridade ao eixo UE-Mercosul', 18 May.

Público (1999b), 'Calçado de sucesso', 8 May.

Santos, H. (1999), 'Projecto Global', *Fortunas e Negócios*, vol.83, February, pp.12-19.

Silva Lopes, J. da (1996), *A Economia Portuguesa desde 1960*, Gradiva, Lisbon.

Wise, P. (1999) 'Venturing abroad in the spirit of the early navigators' *Financial Times*, Portuguese Banking and Finance, 31 March, pp.11-13.

3 Economic Change and Regional Development in Portugal

STEPHEN SYRETT

Introduction

Over the last 20 years the Portuguese economy experienced rapid economic modernisation which had profound effects on the country's regional geography. Integral to the changed sectoral composition and macroeconomic performance of the national Portuguese economy was a changed spatial organisation of production. Economic transition rooted within existing patterns of uneven development was characterised by the restructuring of traditional industrial centres, the growth of new centres of economic activity and the changed insertion of traditional marginal rural areas. The result was fundamental change in the nature of regions and localities throughout Portugal and in the lived experiences of all; whether they resided in the major metropolitan cities of the coastal strip or the marginal rural areas of the interior.

This chapter analyses the relationship between economic transition and the changed urban and regional system in Portugal. The first section considers the nature of economic modernisation and growth within Portugal and focuses on the profound sectoral changes experienced over the last 25 years. The second section explores the relationships between sectoral change and uneven regional development, both at the national level and through the analysis of a number of selected regional case examples. This discussion demonstrates how the interrelationship between processes of economic change and specific regional contexts produced very different regional dynamics. The final part of the chapter examines the policy issues raised by these changes and identifies the principal challenges for local and regional development policy in Portugal at the start of the twenty-first century.

Economic Modernisation and Growth in Portugal

In the post-revolutionary period the Portuguese economy underwent fundamental structural change. Although the period 1974-85 was one of considerable economic instability, characterised by high inflation, a large public sector debt, rising unemployment and unstable political conditions, the period after 1986 was one of sustained economic growth. From accession to the EU (European Union) in 1986, Portugal consistently displayed some of the highest economic growth rates within the EU. It also achieved a significant decline in inflation, interest rates and public debt, alongside improved employment generation. The economy grew particularly strongly in the late 1980s and achieved average annual growth rates of 4.6% between 1985-91. After an economic downturn in the early 1990s, growth returned in the mid-1990s to achieve annual rates of over 3% between 1995-2000 (see figure 3.1). Inflation, which remained at high levels throughout the 1980s, fell significantly in the 1990s to levels of below 3% (see Figure 3.2). Unemployment, traditionally low by EU standards, fell to its lowest levels in the early 1990s and despite rising to a high of 7.3% in 1996 following the recession of the early 1990s, fell to 4.1% by 2000.

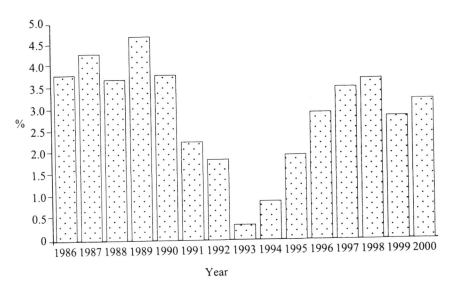

Source: OECD

Figure 3.1 GDP Growth (% annual change), 1986-2000

The apparent strong performance of the Portuguese economy over the last 15 years of the twentieth century reflected a number of factors. First, EU Structural Funds promoted large-scale investment in modernising industrial and physical infrastructures. Second, national state policies promoted a major liberalisation of the economy through a substantial privatisation programme, financial deregulation and the reduction of workers rights. Third, EU membership, low labour costs and increasing economic liberalisation, provided the basis for increased foreign direct investment. Fourth, rising domestic levels of wealth and income and a shift towards a more modern, consumer based lifestyle, stimulated the growth of existing and new domestic markets particularly related to service based activities. Fifth, political stability derived from a series of majority governments, a maturing democracy, and integration into the EU, provided a strong degree of policy continuity. This improved economic performance, in combination with a variety of fiscal reforms and tight budgetary control, enabled Portugal to become a founder member of the European single currency in 1999.

However it is necessary to remain cautious concerning the claims of a Portuguese 'economic miracle'. Despite the strong recent performance on

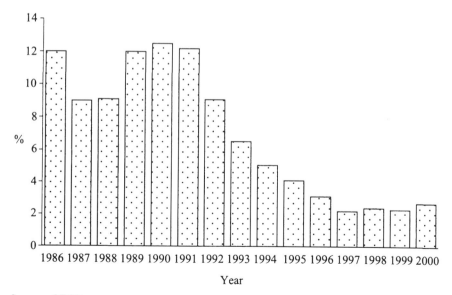

Source: OECD

Figure 3.2 Inflation (% annual change), 1986-2000

certain economic indicators, major structural problems persist. Many sectors continue to rely heavily on low cost labour and resource availability to remain competitive and exhibit continued low levels of investment, technology and training. The majority of Portugal's numerous small firms are characterised by poor management, a weak financial position, low productivity and poorly skilled workers. Agricultural productivity remains low necessitating large amounts of food imports, and the development of producer services still remains weak. The export base exhibits continued reliance on a small number of traditional industries, many of which are either labour intensive or resource based (e.g. textiles and clothing, wood products, minerals, leather, paper and paste) (Corkill, 1999; see Corkill, Chapter 2).

Dimensions of Sectoral Change

Economic modernisation resulted in a fundamental shift in the basic structure of the Portuguese economy (see Table 3.1) and convergence towards the sectoral composition characteristic of the advanced industrial nations of Europe. A long-term consistent decline in the traditional importance of agriculture and resource-based primary industry reduced employment in this sector from 44% of the total workforce in 1960 to 13% by 2000. However this decline levelled out in the latter part of the 1990s and by EU standards this remains a large primary sector workforce. Despite relatively high levels of employment, this sector provided only 4.0% of national gross added value.

  Consistent low levels of productivity in Portuguese agriculture are rooted in the heavily protected, traditional style of agricultural production

**Table 3.1 Change in Sectoral Distribution of Employment (%),
1975-95**

	1975	1985	1995
Agriculture	33.9	21.5	11.5
Industry	33.8	33.9	32.2
Services	32.2	44.5	56.3

Source: Eurostat, 1981 & 1996

which characterised the *Estado Novo*. European integration and the subsequent removal of protectionist barriers led to the contraction of the sector. Despite large outflows of workers from rural areas, agricultural productivity remained low in consequence of the ageing and poorly trained agricultural workforce, poor soils and climate in many regions, small farm sizes, and low levels of investment. As a result Portuguese agriculture struggled to compete in more liberalised food markets. Attempts to modernise Portuguese agriculture, which included major EU-supported programmes (e.g. PEDAP – *Programa Específico de Desenvolvimento da Agricultura Portuguesa*), achieved limited success in improving the commercialisation of sectors such as wine, fruit and horticulture. Forestry and related products remained important areas of economic activity, notably in terms of exports such as paper, pulp, cork and wood manufactures, although the traditionally important fishing industry declined rapidly as a result of competition and international programmes to reduce capacity.

After the rapid industrialisation of the 1960-73 period when the manufacturing sector grew to account for 33% of the workforce (Baklanoff, 1978), manufacturing and construction employment remained relatively stable, rising to a high point of 36% in 1982 and returning to a level of 35% by 2000. However, despite the apparent stability of these employment levels, the manufacturing sector experienced major structural changes (Lopes, 1996). Across the 1980s and 1990s a number of indigenous industries lost significant numbers of jobs, although this was partially offset by new inward investment from multinational corporations (MNCs) (Gaspar & Williams, 1991). As traditional national state protection policies weakened, Portuguese firms became increasingly integrated into more liberalised domestic, European and global markets, and faced new competition from lower-cost producers from Eastern Europe and developing countries, and from better quality products from within the EU (Smallbone et al, 1999). Such competition placed pressure on traditional indigenous manufacturing industry with its competitive base rooted in low labour costs, flexible labour markets and resource availability. The resultant pressures to increase productivity and the subsequent restructuring process often led to significant sectoral job losses, for example, the textiles, clothing and footwear sectors lost 25,000 jobs in the early 1990s.

Processes of industrial modernisation were highly uneven between sectors and firm types (Syrett, 1996). In certain sectors, restructuring was helped by the influx of foreign investment (for example in the food industry, glass and footwear), plus the availability of subsidies for industrial modernisation via the EU-funded industrial modernisation programme (PEDIP – *Programa Específico de Desenvolvimento da Indústria Portuguesa*). A small number of emerging sectors (machinery and electrical products, cars and car components) witnessed growing levels of investment, productivity and employment. Particularly significant was the US $2,099m. investment by Ford and Volkswagen in the AutoEuropa operation in Palmela, south of Lisbon, in the mid-1990s which stimulated the growth of the car-component industry to employ 34,600 workers and export 66% of production. However in other sectors adjustment was less successful and restricted to isolated firms, with the majority of SMEs still beset by problems of a lack of investment, weak management and a poorly trained workforce.

The construction industry grew strongly from the mid 1980s, stimulated by economic growth, rapid urbanisation and heavy investment in physical infrastructure. At the end of the 1990s it experienced an unprecedented boom and by 2000 construction employed 11.0% of the workforce, contributed 13.0% to GDP and was responsible for nearly 50% of total investment. The rapid employment growth of the mid 1990s arose from a number of large public sector construction projects (for example the Vasco de Gama bridge across the Tagus, the extension of the Lisbon underground, the EXPO 98 site in Lisbon, and the train connection across the Tagus), alongside a housing construction boom stimulated by low interest rates. The resultant demand for construction labour sponsored the growth of the immigrant workforce, particularly in Lisbon (see Eaton, Chapter 5). The domestic housing boom began to falter in 2000 and, in the medium term, reductions in EU funds will lessen the scale of public sector construction investment. However, in the short term, the sector appears set to remain important due to existing physical investment plans for the period to 2006 and other large projects, such as a new Lisbon airport at Orta, and the stadia for the Euro2004 football tournament.

The long term trend towards the tertiarisation of the Portuguese economy meant that by 2000 the services sector accounted for 52% of employment and 66% of national gross added value. This tertiarisation process was characterised by a number of dimensions. First, the development of telecommunications, banking and insurance and producer services related to economic modernisation and liberalisation. Second, the continued growth of tourist industries as a result of expanding international visitor numbers and to a lesser extent via the growth of domestic tourism (see Williams, Chapter 4). Third, the evolution of welfare provision, particularly in health and education, and various realms of public administration as the national and sub-national state expanded the scope and scale of its activities. Fourth, the growth of consumer and retail services related to changing lifestyles and higher levels of personal wealth which have resulted in the rapid growth of hypermarkets and shopping centres.

Fundamental to the growth of the high profile service-sector industries that witnessed the highest productivity and employment growth (e.g. banking, producer services, telecommunications and information technology) was the shift towards deregulation and liberalisation, often linked to privatisation programmes that attracted foreign and domestic investment. For example the moribund state-owned banking sector was transformed via limited liberalisation in the 1980s and privatisation in the 1990s. This restructuring process resulted in a massive growth in productivity, as well as

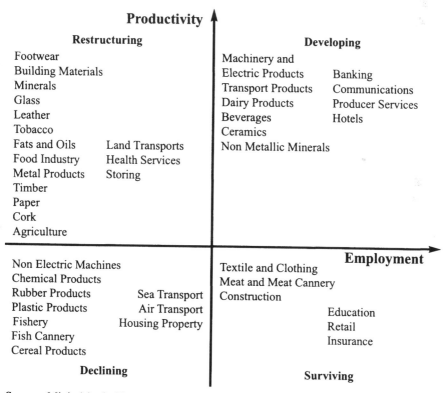

Source: Ministério do Plano e Administração do Território, 1993

Figure 3.3 Patterns of Sectoral Change, 1986-92

expansion of the banking network (from 0.15 branches per 1,000 inhabitants in 1985 to 0.41 per 1,000 in 1997) and the availability of financial products. Similarly in the telecommunications industry, the privatisation of the state monopoly company, sponsored rapid sectoral growth (see Sousa, Chapter 6) with a massive expansion of the fixed line and particularly the mobile phone network. As a result, in 1998 this sector accounted for 5.3% of GDP and 4.0% of investment, although it contributed only 0.5% of employment.

Examination of the overall process of sectoral transition reveals significant differences in the relative performance of subsectors. For example, MPAT's (1993) analysis of sectoral performance between 1986-92, in terms of changes in employment and productivity, identified the following groups (see Figure 3.3):

- *Developing: growth in employment and productivity* This group comprised a range of high profile service industries (e.g. banking, producer services, hotels, telecommunications) and a number of manufacturing sectors related to growth in domestic and export markets (machinery and electric products, transport products, diary products, beverages, ceramics and non-metallic minerals). Central to the growth dynamic of these sectors was favourable market conditions and increased levels of foreign and domestic investment often related to privatisation programmes, as well as, in certain cases, the availability of EU funding.
- *Restructuring: growth in productivity but declining employment* A group which comprised more traditional economic sectors (e.g. footwear, glass, minerals, leather, building materials, food industry, fats and oils, metal products timber, pulp, paper cork agriculture, land transport, health services, storage) which had begun to restructure activity in order to increase productivity and cope with increased competition from both domestic and foreign firms. This restructuring process tended to lead to job losses and was stimulated by higher levels of foreign investment (e.g. in the food industry, pulp and paper, glass and footwear), and the availability of subsidies for industrial modernisation (via PEDIP) or for agricultural modernisation (via PEDAP).
- *Surviving: some employment growth but continuing low levels of productivity* This group consisted primarily of service industries (e.g. education, retail and insurance) related to the growth of domestic consumption and construction. In the 1980s it also included some more traditional, labour intensive industrial activity (textiles and

clothing) and food processing (meat and meat canning), which during this period grew on a competitive basis of cheap labour but subsequently lost employment later in the 1990s.

- *Declining: declining employment and low levels of productivity* This group was characterised by sectors which were slow to modernise and operated in markets which were either declining and/or had become highly competitive internationally (e.g. chemical, rubber and plastic products, fishery related activity, non-electric machines, sea and air transport, housing property, cereal products). A number of these sectors had been traditionally highly protected domestically (e.g. fisheries, cereal production, air transport) and producers were slow to respond to increased competition and changes in international regulation (e.g. fisheries, air transport).

The highly uneven nature of sectoral transition within Portugal demonstrated by such analysis points to the importance of looking behind the macroeconomic indicators to question the solidity of the Portuguese economic success story. Examination of sectoral performance indicates continued structural weaknesses in many of Portugal's key industrial sectors and the highly variable nature of firm response to rapidly changed market conditions. Whilst certain selected sectors witnessed employment and productivity growth as a result of programmes of liberalisation, modernisation and privatisation, others experienced dramatic job losses and worsened conditions of pay and employment. A more liberalised economy resulted in economic growth, but also impacted upon labour markets to increase disparities and social polarisation both within and between sectors. These trends are intimately related to the changing spatial basis of production within Portugal and hence have sponsored important changes to processes of uneven development.

Regional Impacts of Economic Change

The traditional starting point for describing the economic geography of Portugal is in terms of a coastal-interior divide (Lewis and Williams, 1981). This spatial divide remains strongly in evidence and was in certain respects reinforced by the processes of socio-economic change that took place over the last quarter of the twentieth century. As Figure 3.4 demonstrates, the vast majority of the Portuguese population remains concentrated in the relatively

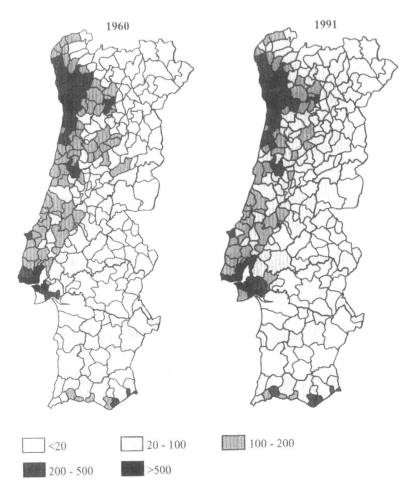

Source: Instituto Nacional de Estatística

Figure 3.4 Population Density by Concelho, 1960 and 1991

narrow coastal belt (or littoral) which extends from Viana de Castelo in the north to the Setúbal peninsula in the South (INE, 1993). Although it is evident that all regions of Portugal benefited from improvements in basic levels of human development such as life expectancy, literacy rates, access to basic infrastructures and per capita wealth, significant regional disparities remain (see Figure 3.6). The coastal strip, including the major metropolitan areas of Lisbon and Porto, and the Algarve in the South, continues to enjoy

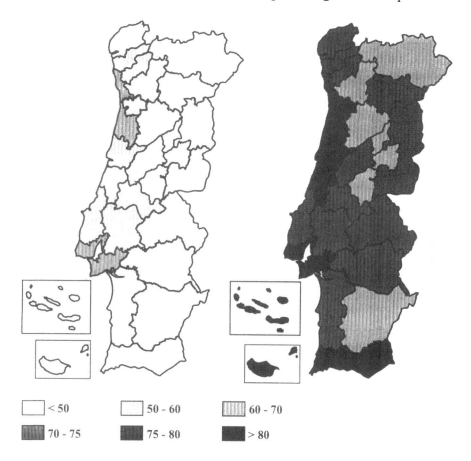

(*) The Composite Index of Human Development combines four components each measured on a scale of 1 to 100:

life expectancy (85 years = 100);

literacy (all population can read and write = 100);

comfort (all population have piped water, electricity and sanitation = 100);

GDP per capita (highest GDP level for a local authority area in 1997 in constant prices = 100)

Source: Minstério do Planeamento, 1999, p.I-70

Figure 3.5 Composite Human Development Index(*), 1970 and 1997 (NUTS III)

much higher levels of socio-economic development, a difference strongly rooted in the enduring uneven spatial distribution of economic wealth.

Processes of socio-economic change not only reinforced existing spatial divisions but also sponsored new patterns of local and regional development.

This is particularly notable with regard to rapid urbanisation, evident not just in the large metropolitan areas but throughout Portugal. Allied to economic growth, urbanisation processes intensified markedly from the 1960s onwards (see Figure 3.6) (Ferrão, 1996). Central to this urban growth were the metropolitan areas of Lisbon and Porto, which grew to account for over 35% of the Portuguese population. Whilst these two cities continue to dominate Portugal's urban system, population growth in the Metropolitan Areas of Lisbon and Porto stabilised across the 1980s (see Figure 3.7).

Intensified urbanisation in the 1990s led to some dispersal of the urban system, with highest population growth rates notable in the areas surrounding the large metropolitan areas as well as in many of the medium sized cities both within and outside the coastal belt. For example in the wider Lisbon region, between 1991-2001 the most dramatic increases in the resident population were found in towns such as Sintra (39.3%), Sesimbra (35.2%), Mafra (24.1%) and Palmela (21.4%). Similar strong population growth

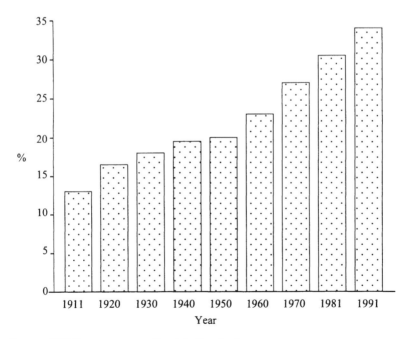

Source: INE, Recenseamento Geral da População

Figure 3.6 Population Resident in Agglomerations of over 10,000 Inhabitants (%), 1911-91

across this period was evident in other medium sized towns such as Leiria (16.5%), Faro (12.6%), Viseu (11.6%) and Guarda (12.9%).[1] These changed patterns of urban development are significant not only in their relationship to the changed geography of economic activity, but also in their profound and complex impact upon the social and cultural fabric of Portugal (Gaspar and Jensen-Butler, 1992). Shifts towards smaller nuclear families and changed class structures, cultural habits and consumption norms both reflect and constitute this transition towards a predominantly urbanised society (Barreto. 1996).

The processes of economic sectoral change discussed previously reinforced the primacy of the coastal belt but also changed the nature of local and regional dynamics (see Figure 3.8) (Costa and Costa, 1996). The coastal

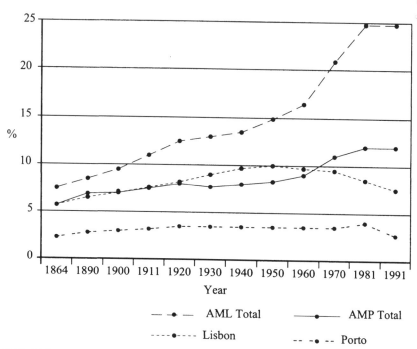

AML: Lisbon Metropolitan Area
AMP: Porto Metropolitan Area

Source: Ferrão, 1996

**Figure 3.7 Population Resident in Metropolitan Areas
of Lisbon and Porto (%), 1864-1991**

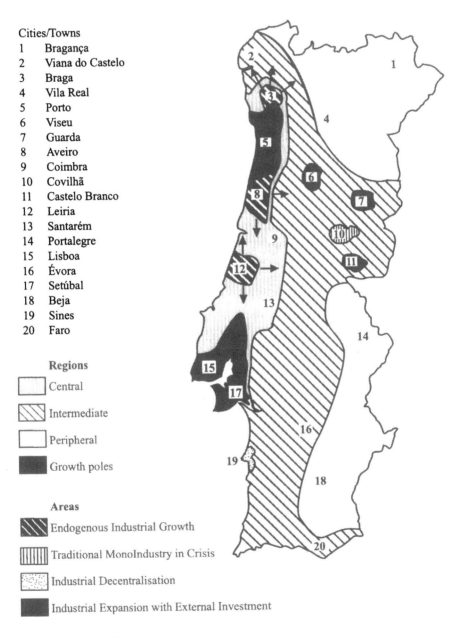

Cities/Towns
1   Bragança
2   Viana do Castelo
3   Braga
4   Vila Real
5   Porto
6   Viseu
7   Guarda
8   Aveiro
9   Coimbra
10   Covilhã
11   Castelo Branco
12   Leiria
13   Santarém
14   Portalegre
15   Lisboa
16   Évora
17   Setúbal
18   Beja
19   Sines
20   Faro

Regions

☐ Central

▨ Intermediate

☐ Peripheral

■ Growth poles

Areas

▧ Endogenous Industrial Growth

▥ Traditional MonoIndustry in Crisis

▦ Industrial Decentralisation

■ Industrial Expansion with External Investment

Source: Caetano, 1995

Figure 3.8 Spatial Economic Development in Portugal

belt retains its role as the focus of manufacturing activity (accounting for 83% of manufacturing firms, 92% employment and 94% turnover). Given that the distribution of service sector activity closely follows that of population, the littoral is also the focus of the most rapidly growing service sector industries. Lisbon and its surrounding area remains the predominant economic centre in terms of its wide range of service and manufacturing activity, with the Porto Metropolitan Area the second industrial centre, particularly with regard to manufacturing production of consumer goods. Elsewhere in the coastal belt a defining feature is a range of geographically focused industrial clusters normally specialised in one sector (e.g. textiles and clothing in Braga, footwear in S. João de Madeira), but also via inter-industry linkages across diverse sectors (e.g. metal products, electrical goods, wood products in Águeda) (Caetano, 1995; Ferrão and Mendes Baptista, 1992).

Outside of the coastal belt an intermediate region is characterised by relatively low levels of economic activity with economic growth focused in large towns (district capitals such as Viseu, Guarda, Castelo Branco) which benefited from increased accessibility, the growth of commercial activities and some industrial development often via inward investment. This area also includes declining industrial areas (e.g. the woollen industry of Covilhã and Serra de Estrela and capital intensive heavy industries at Sines). The 'peripheral' interior region remains strongly rural with little industrial activity. The area is characterised by marginalised and often stagnant agricultural related activity, although there are examples of more commercially oriented horticultural production alongside some growth of rural tourism and craft (*artesanato*) activities.

Analysis of GDP and population change at the NUTS level III for the 1981-1991 period reveals something of the dynamism of recent regional change (MPAT, 1993) (see Figure 3.9).[2] This analysis identified the differential performance of regions as follows:

- *Developing regions: (growth in population and GDP levels):* This group included areas closely linked to the expansion of the traditional core regions of Lisbon and Porto (Tamega, Cavado, Douro Vouga, Pinhal Littoral, Oeste). These were often the focus for a variety of industrial districts and benefited from improved transport infrastructures. This group also included areas which developed through specialisation in tourism (Algarve, Madeira).
- *Surviving regions (growth in population but some decline in GDP level):* These comprised traditional industrial centres of activity

(e.g. Lisbon, Porto, Vale de Ave, Setúbal, Baixo Vouga). These regions were forced to adjust to increasing levels of competition in either export (clothing) or domestic markets (e.g. electronics), or the restructuring of older capital intensive industries (pharmaceuticals, shipbuilding). These regions benefited from major inflows of foreign capital to help restructure the productive base, as well as considerable investment of state and EU resources to help modernise Portugal's economic structures (e.g. Setúbal peninsula).

- *Restructuring regions (some population decline, but rising GDP levels):* These interior regions experienced some levels of economic growth (Minho-Lima,Dão-Lafões, Beira Interior Sul, Alto Alentejo, Alentejo Central, Alentejo Litoral) often related to improved accessibility and increased levels of East-West flows of commercial traffic. Growth tended to focus within the major urban centres (e.g. Viseu, Castelo Branco, Évora) stimulated by inward investment and the growth of regional service activity.

- *Abandoned regions (large population decline but rising GDP levels):* These regions were characterised by very narrow economic bases, predominantly agricultural with some low productivity commercial activity (e.g. Pinhal Sul, Trás-os-Montes, Baixo Alentejo, Pinhal Norte, Beira Norte). They suffered ongoing rural depopulation, low levels of educational attainment and an ageing population. The rise in GDP levels reflected the very low starting base to GDP levels, isolated investments (e.g. forestry in Pinhal Sul; mining in Baixo Alentejo) and a significant rise in state and EU spending within these regions.

- *Dipping regions (population decline and stagnant or falling GDP levels):* These regions were characterised by agricultural specialisation and operation in sectors of increasing competition (e.g. Douro wine production, Cova da Beira fruit and horticultural production; Serra de Estrela pastoral farming, cheese production and declining woollen industry).

- *Decreasing regions (minor population decline and stagnant or slightly falling GDP levels):* These comprised regions in or proximate to the Littoral (Medio Tejo, Baixo Mondego, Lezíria) which were dependent on primary sectors and limited processing activity and which were negatively affected by technical change and rationalisation (e.g. pulp, food and paper in Baixo Mondego). The islands of the Azores and their reliance on diary production, also fall within this group.

Whilst such categorisations are crude and not unproblematic - particularly

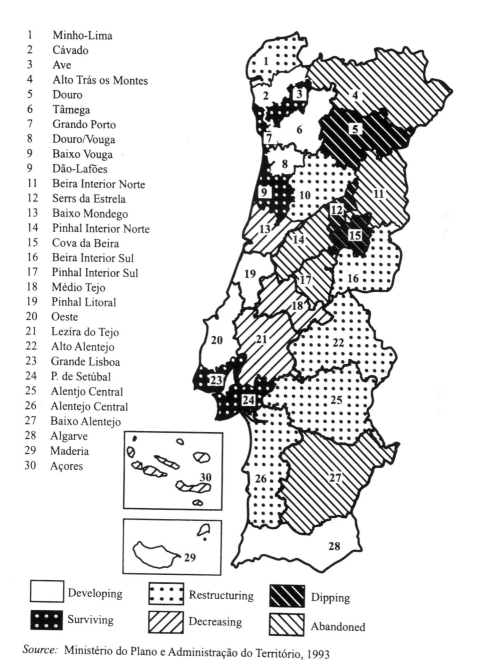

1   Minho-Lima
2   Cávado
3   Ave
4   Alto Trás os Montes
5   Douro
6   Tâmega
7   Grando Porto
8   Douro/Vouga
9   Baixo Vouga
9   Dão-Lafões
11 Beira Interior Norte
12 Serrs da Estrela
13 Baixo Mondego
14 Pinhal Interior Norte
15 Cova da Beira
16 Beira Interior Sul
17 Pinhal Interior Sul
18 Médio Tejo
19 Pinhal Litoral
20 Oeste
21 Lezíra do Tejo
22 Alto Alentejo
23 Grande Lisboa
24 P. de Setúbal
25 Alentjo Central
26 Alentejo Central
27 Baixo Alentejo
28 Algarve
29 Maderia
30 Açores

Developing   Restructuring   Dipping

Surviving   Decreasing   Abandoned

Source: Ministério do Plano e Administração do Território, 1993

Figure 3.9 Change in Portuguese Regions, 1981-91

with regard to the quality of the secondary data on which they are based - they do provide insights into the dynamism of a recent period of regional change. Those regions that either performed well or demonstrated a degree of resilience to change were located predominately in the coastal belt and the Algarve, although the emergence of a number of more accessible interior regions was also evident. The regions that performed less well were located in the Interior, although with some coastal areas also present, and tended to be predominately rural, agricultural areas which experienced population loss and difficulties in improving levels of productivity.

In order to understand this complex relationship of socio-economic change and local/regional development in more detail, it is valuable to consider some case examples that demonstrate aspects of economic restructuring in more detail.

Case Studies: Regions of Growth, Restructuring and Decline

Lisbon Metropolitan Area (AML)

By the end of the 1990s, GDP per capita levels in the Lisbon and Tagus valley region stood at 89 against an EU average of 100; a substantial rise from 76.6 for the 1989-91 period. This successful convergence towards EU average GDP per capita levels means that after a transitory period between 2000-2006, this region will lose its Objective One eligibility for EU Structural Funds.[3] This region is traditionally the largest and most productive manufacturing region in Portugal. The Lisbon and Tagus Valley area in 1992 accounted for 29% of manufacturing jobs and 51% of the domestic industrial gross added value. However, service industries have become increasingly important within the region. Across the 1981-91 period, the region witnessed a rise in tertiary employment from 61% of the working population to 70%, and a fall in secondary sector employment from 36% to 28%. Service sector growth is associated with the growth of producer services, tourism and consumer services, reflecting Lisbon's role as a major European city and the increasing wealth of certain sections of the population.

In the 1980s this area underwent a major restructuring of its manufacturing base (see Vale, 1998), which led to large increases in unemployment including rates of over 20% on the Setúbal peninsula. In the 1990s unemployment levels fell and there was evidence of dynamism in terms of rising levels of workers qualifications and increased exports from some sectors.

Traditional industrial sectors such as textiles and clothing, chemicals and iron and steel witnessed large job losses, however other manufacturing sectors suffered only small employment losses (e.g. metal products, non-metal products and equipment) and others even witnessed a small increase (e.g. food and beverages) (see Table 3.2).

The region has been the major beneficiary of flows of foreign direct investment. In the 1988-92 period, Lisbon and the Tagus valley received 85% of total foreign direct investment, with the AML accounting for the vast majority of this. Large flows of foreign investment into sectors such as banking and finance (30.3% of total FDI flows 1988-92) and real estate and producer services (25.4%) principally benefited the Lisbon region. The area was also the recipient of the single largest FDI investment in Portugal, the Ford/VW works (AutoEuropa) at Palmela, on the Setúbal peninsula. In addition to the flows of private investment, the region was a major beneficiary of European Structural Funds. This included both support for

Table 3.2 Manufacturing Employment Change in the Lisbon Metropolitan Area, 1988-93

	Employment (1993)	% of Employment (1993)	Employment change 1988-93 (%)
Food, beverages and tobacco	26,252	14.98	3
Textiles, clothing and leather	18,247	10.41	-21
Wood and cork products	8,685	4.96	-2
Paper, pulp, printing and publishing	20,749	11.84	20
Chemicals	22,662	12.93	-29
Non-metal products	10,254	5.85	-3
Iron and steel	4,416	2.52	-40
Metal products, electronics and transport	62,027	35.40	-4
Other industries	1,947	1.11	–
Total	175,239	100.00	-9

Source: Vale (1998)

manufacturing restructuring (e.g. via PEDIP and an EU funded Integrated Development Operation on the Setúbal peninsula) and major public infrastructure projects (e.g. the new Tagus bridge, the Lisbon underground, airport, rail system, motorway system etc.). The result of this was a major construction boom in the 1990s and improved accessibility, which strengthened Lisbon's role as the principal nucleus of tertiary activity.

These changes had important spatial variations within Lisbon. Tertiary activity remained focused in the Lisbon municipality itself, although some decentralisation has occurred. Tertiary employment in Lisbon municipality rose from 74% to 80% of employment in the 1981-91 period, whilst manufacturing employment fell by over 20% in the 1988-93 period. Some decentralisation of manufacturing activity is also evident with the highest growth rates in manufacturing employment of over 20% located in the areas of Palmela, Sesimbra and Sintra. Rapid economic change and development also saw the emergence of increased social polarisation and spatial segregation within Lisbon (Gaspar, 1997). Pockets of social deprivation are now increasingly visible, characterised by factors such as poor public services, high levels of unemployment, drug abuse, lack of housing provision, and immigrant populations (Silva, 1999).

Marinha Grande

Marinha Grande is an industrial district located close to Leiria in the centre of the coastal belt; an area which has experienced relatively high rates of GDP and population growth particularly around towns such as Leiria and Pombal. Marinha Grande itself has a unique history. It is one of Portugal's oldest industrial centres with a long tradition of glass production. As a result the locality is dominated by industrial employment which has historically led to a traditionally strong left wing political culture and an important role for organised labour. Over the last 20 years this area witnessed dramatic transition in its industrial structure. The traditional glass industry underwent major restructuring and many aspects continue to experience considerable problems. At the same time, the area witnessed strong growth in the moulds used for the creation of injected plastic mouldings, an industry that initially grew out of expertise for making moulds for the glass industry. The continued growth of the moulds industry provided new dynamism to the area and, to some extent, offset the large job losses from the glass industry. This industrial cluster, based around the glass, moulds, plastics and packaging industries (see Figure 3.10), is typified by intense interfirm relations producing a highly integrated, localised industrial system.

Industrial employment in MG fell by 12% between 1982 and 1993. Across this period employment in the glass industry fell by 35.8% (a loss of 2,050 jobs) with large job losses in the early 1980s. In comparison the mould industry witnessed steady employment growth, particularly from the mid 1980s, to create over a 1,000 new jobs. Population levels remained relatively stable in the 1980s and grew slightly (+ 5.8%) in the 1991-2001 period. However, unemployment persisted due to the difficulties of retraining redundant glass workers for the new jobs created in the moulds industry.

The rapid development of the moulds industry was accompanied by the development of various specialised institutional structures such as training facilities (CENFIM – *Centro de Formação Profissional da Indústria Metalúrgica e Metalomecânica*), a technology centre (CENTINFE – *Centro Tecnológico para a Indústria de Moldes e Ferramentas Especiais*) and an active employers organisation (CEFAMOL – *Associação dos Industriais de Moldes*). In addition a number of producer services organisations, particularly in the areas of design, accountancy and commercial agents, developed to serve the specialised needs of the moulds industry. Due to the demands for skilled personnel for the moulds industry, both in terms of basic workers and

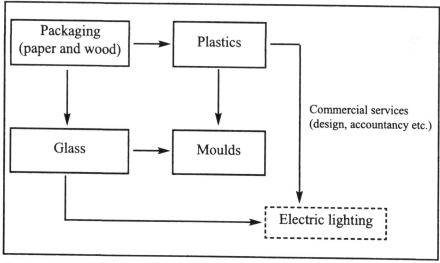

Source: OEFP, 1995

Figure 3.10 The Marinha Grande Industrial Cluster

higher order technical and managerial positions, there was a small but notable increase in the levels of skills within the workforce (*Observatório do Emprego e Formação Profissional*, 1995). Economic change was also accompanied by a changed local political culture. As the large glass factories characterised by well-organised trades unions declined, they were replaced by the small businesses of the moulds industry where trades unions were less powerful or, very often, non-existent. In local politics, after many years of Communist Party dominance, in the 1990s control of the municipality passed to the Socialist Party, and even for a brief period, to the Social Democratic Party (PSD).

The localised growth of the moulds industry provides an interesting insight into the modernisation of Portuguese industry. This sector is dominated by the developed industrial economies as it requires the production of precision metal products (e.g. the moulds) which in turn require high levels of investment in computer controlled machinery and a relatively skilled labour force. The Portuguese moulds industry, focused in the two centres of Marinha Grande and Oliveira de Azemeis, exports 90% of production mainly to the EU and North America (CEFAMOL, 2001), a figure which accounts for over 20% of all Portuguese metal engineering sector exports. That the moulds industry of Marinha Grande is able to compete successfully internationally illustrates the achievement of a certain level of industrial development, technological innovation and workforce skills. Yet despite continued investment, Portuguese productivity levels in this sector remain the lowest of the major producers as do labour costs, indicating the continued importance of low cost labour in underwriting the sector's international competivity.[4]

Marginal Rural Areas: Alto Trás-os-Montes

The rural area of Alto Trás-os-Montes e Douro has suffered a long term decline of population (see Figure 3.11). Between 1960 and 1981 this region lost 138,000 inhabitants, and a further 64,000 inhabitants were lost across the 1980s. The Alto Trás-os-Montes in particular continues to suffer some of the highest population losses via outmigration of any region within Portugal. Between 1981- 1991 the resident population decreased by 15.0%, a loss not just confined to the smaller villages but which, unlike other areas, also included the principal urban centres (e.g. Braganza -9%, Chaves, -13%, Mirandela, -15.9%). The 1991-2001 period saw a slow down in the rate of population decline (-5.2% against a national average increase of +4.9%) (INE, 2001). However in this period the major urban centres experienced small population increases (Chaves +6.4%, Braganza +4.9%, Mirandela +2.4%) whilst all rural municipalities experienced losses, in many cases at very high rates (e.g. -19.2% in Boticas and -17.3% in Montalegre).

This continued outflow of population has left an ageing population. Between 1981-91 the population of the 0-14 age group fell by 35.4% and the 15-24 age group by 26.1%. In contrast the 25-64 age group suffered a much smaller population loss of 3.4% whilst the over 65 population group grew by 15.4% (see Table 3.3). Alongside population loss the area performed poorly on a range of other economic and social indices including low levels of GDP and purchasing power. Unemployment in 1996 was 7.6%, the second highest in the Northern region. The area also has the highest infant mortality rates in Portugal, an average of 14.6 per thousand between 1991-95, much higher than the second highest value for the Azores of 11.6 per thousand.

The productive structure of Alto Trás-os-Montes remains dominated by small enterprises, principally based in the agricultural sector though there was some growth in public and small scale commercial services related to serving the local population and tourist activity. Agricultural employment although still high, has fallen, and agricultural productivity is low, as are incomes in this sector. The area is now more accessible than in the recent past due to improvements in the road networks. However it has not directly benefited from the development of East-West links in the manner that some other interior regions have, and there has been little inflow of investment into the area. More positively the region does have a well distributed urban system which provides a basis for the supply of public services within the region.

A primary challenge for this region is to generate new employment opportunities for young people with rising levels of qualifications, in order to retain them in the region. To this end there have been a number of policy

Table 3.3 Change in Age Structure in Alto Trás-os-Montes, 1981-91

Age Group	1981	1991	% Change
0-14	27.0	20.2	-35.4
15-24	18.0	15.4	-26.1
25-64	42.2	47.3	-3.4
+65	12.8	17.2	+15.4

Source: INE

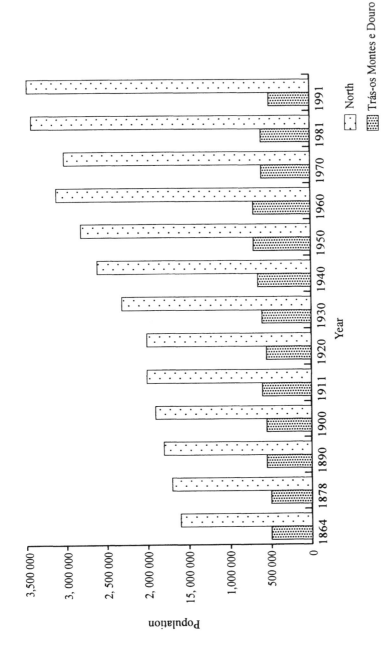

Source: Instituto Nacional de Estatística

Figure 3.11 Population Change in Trás-os-Montes and Alto Douro, 1864–1991

interventions which have sought to commercialise specialised agricultural products (e.g. the development of indigenous cattle breeds) and agro-food products (e.g. smoked meats, chestnuts and oil). Rural tourism has also been promoted focusing on the natural beauty of the area (e.g. Parque Nacional Montesinho. Parque Natural do Alvão, Parque Natural Peneda-Gerês) and the development of the Parque Arqueológico do Côa, famous for its cave drawings. Although there has been a significant expansion in the supply of accommodation (via schemes such as *Turismo de Habitação* and *Turismo Rural*) and the promotion of local gastronomy and craft products, the scope for tourism remains limited by the lack of tourist infrastructures (e.g. hotels, restaurants etc.) and the scale of demand.

Developments in Regional Policy

Traditionally regional policy was a neglected arena of concern under the *Estado Novo* (Simões Lopes, 1983). However, from the mid 1980s there was a significant increase in, and interest in, policies which addressed regional inequality and regional imbalance. Such concern was largely the result of Portugal's accession to the EC. As a major recipient of European regional funds, Portugal was forced to develop national and regional development strategies which met the approval of the European Commission (Syrett, 1997a). The purpose of these EU Structural Funds was to promote real convergence between the Portuguese and Community economies and to promote internal economic and social cohesion.[5] Undoubtedly it is the first of these, the pursuit of national economic growth, which has been the dominant objective, whilst tackling internal regional disparities has been seen very much as a secondary and supportive objective. The majority of European regional funds have been spent on a huge round of investment in basic public infrastructures (roads, water supply, sanitation, railways, airports, hospitals, universities etc.), but these monies have also provided significant levels of funding to support the modernisation of the Portuguese economic structure and improve levels of skills and education.

The estimated impact of the second Community Support Framework (CSFII),[6] which operated across the 1994-99 period, was an increase of 3.3% in real GDP (an average yearly growth rate of around 0.53 percentage points). The CSF II was also credited with the creation of 106,000 jobs (30% in civil construction) and a fall in unemployment of two percentage points (Ministério do Planeamento, 1999). Its biggest impacts were in sponsoring large-scale infrastructure development, whereas impacts in areas such as innovation and R&D were less apparent. With regard to regional disparities, as Table 3.4 illustrates, this period saw the continuation of a trend towards a degree of regional convergence. Since 1989 all regions witnessed per capita GDP growth, but differential growth rates have led to some narrowing of the gap between the richest and poorest regions. However, these figures at the NUTS level II need to be treated cautiously as they mask significant intra-regional variations, particularly within regions that span both coastal and interior areas. Even after this convergence, inter-regional variations in GDP remain significant between Lisbon and the Tagus Valley, the wealthiest region (89.3 against EU average of 100), compared to the poorest mainland region (Alentejo, 61.2) and the islands (Azores, 49.7). Furthermore, these figures confirm that outside of Lisbon and the Tagus Valley, the regions of Portugal continue to have low levels of GDP per capita compared to the EU average.

Given the planned expansion of the EU, it seems likely that the Third Community Support Framework (CSF III, 2000-06) will be the last which involves a high level of EU transfers to all areas of Portugal, thus drawing to a close a unique period which started in the mid 1980s. With the planned accession of a number of poorer nations to the EU, Portugal will lose its high

Table 3.4 Regional Change in Portugal (GDP per capita), 1989-99

	GDP per capita		
	1989 (EU12=100) (pps)	1993 (EU15=100) (pps)	1999 (EU15=100) (pps)
North	50	59.6	65.6
Centre	40	55.2	62.1
Lisbon & Vale de Tejo	77	87.4	89.3
Alentejo	34	54.4	61.2
Algarve	48	70.6	71.4
Azores	–	49.2	49.7
Madeira	–	50.5	59.1
Portugal	57	67.7	72.3

pps = purchasing power standards

Source: CEC, 1990; Ministério do Planeamento, 2000

priority position and future transfers are set to be more limited in scale and more likely to be targeted to Portugal's most disadvantaged areas. The key objective of the CSF III is to focus on improving the competitiveness of the Portuguese economy (Ministério do Planeamento, 2000). This reflects recognition that recent growth was characterised by limited improvements in productivity and that the economy remains poorly prepared to face more intense global competition. Consequently the three priority areas of intervention under CSF III comprise:

- human potential: to improve skill levels in order to increase productivity;
- productive activity: to achieve growth in competitiveness via support for business strategies, R&D and innovation and producer services, as well as support for agriculture, rural development and fisheries;
- territorial organisation: to pursue infrastructural improvements in a manner that is compatible with environmental and regional development needs and in order to reduce regional disparities.

Table 3.5 illustrates how these were translated into four priorities within the CSF III, financed by a total budget of over 42 billion Euros, 48% of which comes from the EU Structural Funds, 29% from national public resources and 22% from private finance.

Challenges for Urban and Regional Policy

The Community Support Frameworks, and their forerunners, have been highly important in guiding policies promoting national economic development and in supplying the majority of resources for regional and local development programmes. The CSFs naturally reflect the European Commission's own guideline priorities and illustrate how influential EU policy directions have been in setting the Portuguese urban and regional policy agenda. The third CSF, like its predecessors, focuses primarily on the pursuit of national economic development, with regional, local and urban dimensions subsumed beneath this overriding concern. There consequently remains justifiable concern as to whether this programme will effectively address the key regional and local development challenges that Portugal faces at the beginning of the twenty-first century. On the one hand this includes dealing with the stresses of sustained economic growth, particularly with regard to the environment and issues of social polarisation. On the other,

Table 3.5 Community Support Framework III, 2000-06

Main Priorities	Total Cost	Public Expenditure			Private Financing	Cohesion Fund	Other Financial Instruments	European Investment Bank
		Total	Structural Funds	National Public Resources				
Priority 1	7,095	6 728	4 267	2 462	366	–	–	–
Priority 2	13,110	6 094	4 132	1 963	7 015	–	–	85
Priority 3	3,824	3 515	1 721	1 794	309	3 191	16	–
Priority 4	15,275	14 187	8 978	5 209	1 088	108	1	1 234
Total CSF	42,200	32 800	20 535	12 265	9 400	3 299	17	1 319
Reserves	2,896	2 276	1 437	882	622	–	–	–

Euro (millions) current prices

Priorities

1   Improving skills among the Portuguese, and promoting employment and social cohesion
2   Modifying the profile of production towards activities of the future
3   Asserting the value of the land and Portugal's geo-economic position
4   Promoting sustainable regional development and national cohesion

Source: Ministério do Planeamento, 2000

it comprises the need to better target problem areas whether this is at the scale of interior rural regions, older industrial localities or urban neighbourhoods.

Economic Growth Regions: Planning for Growth

Areas which have experienced rapid economic growth (e.g. Lisbon and surrounding areas, the Algarve, localities in the coastal belt and certain key towns in the interior) face some of the negative externalities associated with such growth. Increased congestion, environmental problems, conflicts over land use, rising land values and skills shortages are problems which can strike at the heart of local dynamism, whether in the form of environmental problems in tourist areas, or skills shortages in industrial districts.

A key part of a response to these challenges is the development of a stronger planning framework to manage growth pressures. Past urban growth in Portugal has often been unplanned or poorly regulated (Portas et al, 1998). Newly built residential areas have often lacked adequate social and physical infrastructures, whilst industrial land use has often been mixed with residential and other land uses. More positively, the Portuguese planning system developed significantly over the 1990s with all local authorities now required to produce a PDM (*Plano de Director Municipal*) and new legislation (e.g. Lei 48/98 on spatial and urban planning) strengthened and clarified aspects of the planning system. Yet doubts remain as to whether these developments will impact at the local level. For the planning process to become more effective a wider range of actors need to be actively engaged and local authorities need to have the power - not just legally, but also financially, politically and administratively - to implement planning policies aimed at the creation of long term, sustainable local economic growth.

In addition to a stronger physical planning process, these localities also need to ensure they retain and develop the capacity for future growth. This requires localities to be aware of sectoral trends and have the institutional ability to anticipate and respond to such changes; the attributes of so-called 'intelligent' or 'learning regions'. As the more successful Portuguese industries seek to modernise and move up the value chain, a common need in growth regions is to ensure the development of local labour markets with an appropriate level of education and skills. To achieve this requires not only the ongoing development of national education and training policies, but also locally based education and training provision achieved through close co-operation between the public, private and voluntary sectors and oriented to particular local needs (Syrett, 1997b). Future growth strategies also need to prioritise environmental issues in a manner that has not occurred to date. The full environmental costs of rapid economic growth have only recently been recognised, and a meaningful engagement with notions of sustainable development is required to prevent escalating environmental degradation.

Problem Regions: Reducing Regional Inequalities

There are a variety of less prosperous regions and localities within Portugal, but three broad types, which present rather different policy challenges, appear particularly important:

Interior rural regions: These regions continue to suffer from a weak economic base that makes it difficult to create employment and retain young people. There are possibilities for these regions to restructure their economic base, principally through the modernisation and specialisation of agricultural and forestry production, the development of agro-food industries, rural tourism, and the commercialisation of traditional craft products. One encouraging development over the last decade was the emergence of a network of rural development groups that provide a new and important institutional component for the pursuit of locally based rural development.

Ultimately, whilst these regions may be able to develop certain local economic strategies, their long term survival appears dependent on significant transfers of national state and EU resources. Somewhat surprisingly, although Portuguese agriculture has received considerable support via the EU Structural Funds, it has not been a major beneficiary of the Common Agricultural Policy (CAP). Despite being among the poorest in Europe, Portuguese farmers receive relatively little under the CAP system of production subsidies which discriminate in favour of products (e.g. meat, grain and diary) where Portugal is least competitive.[7] It is not surprising therefore that Portugal has been at the forefront of proposals for the reform of the CAP, proposing a move away from a system that encourages intensive agricultural production towards one that emphasises the social role of agricultural in improving environmental sustainability, landscape protection, product quality and job creation (Financial Times, 2001). Nationally, rural Portugal offers not only important amenity value but also a central cultural role in sustaining many of the traditions (music, dance, food) which remain so important to Portuguese culture. However, the question remains to what extent there exists a political will to transfer public monies to these regions and support a declining rural population, given other competing priorities.

Traditional industrial areas undergoing restructuring: For these areas, the principal challenge is to modernise and transform existing industrial capacity in the face of increased international competition, both from EU countries via the Single Market, and globally from low cost producers in the developing world. Many firms operating within Portugal's traditional industrial sectors have been slow to respond to a more competitive environment and many have continued to enjoy a degree of protection via various transitional arrangements. Ultimately, these areas of traditional production need to undergo an industrial transformation in relation to investment in new technologies, improved skill levels, better product design and quality, which will together increase productivity and make their products more competitive in domestic and export markets. Development strategies in these localities require both the restructuring of traditional industries and attempts to diversify the local economic base. To achieve this requires the strengthening of a range of local institutions (educational and training facilities, development agencies, sectorally based organisations etc.) and greater interfirm co-operation and collaboration.

Urban areas: The rapid urbanisation of Portugal has intensified the existence of a range of social problems often concentrated in particular neighbourhoods within urban areas (Portas et al, 1998). Issues relating to poverty, increased levels of drug use and criminality, and the growth of immigrant populations are creating significant problems of social exclusion within Portugal's urban environment. Traditionally, these areas of urban social policy have not been well developed within the Portuguese context, partly because these are newly emerging social problems associated with Portugal's transition towards a modern, urbanised society. In this respect Portugal has much it can learn from the advanced industrial economies which have long histories of urban policy interventions. Yet the emergence of an urban policy agenda must avoid simplistic policy transfer and be closely related to the specific needs of Portuguese urban areas. A new urban policy agenda requires not only national government input but also levels of local level mobilisation and participation which have not been evident in the past. The successful development of community based development strategies requires the active engagement of a wide range of urban based social groups, as well as strengthening the scope for action by local authorities, parish councils (*juntas de freguesias*), and city wide authorities.

Conclusions

Changed economic and spatial structures are a defining characteristic of Portugal's recent historical development. Increased integration into the European and global economic system has driven forward rapid changes in the productive system and in regional and urban structures. This remodelling of the economic landscape has led to increased specialisation of spatial structures and increased polarisation between rapidly developing urban based localities and marginalised and redundant rural ones. Such change has created a series of new challenges for regional and urban policy. If the less advantaged regions and localities of Portugal are to benefit from processes of national economic development there is therefore a clear need for a reformed and more active set of spatial policies.

The nature of future regional and local policies must reflect both recognition of the rapidly changing external environment and the strengths of existing forms of social, economic and cultural organisation. As well as the necessity of strengthening the competitive basis of local and regional economies so they can compete internationally, the policy agenda must not lose sight of the enduring place of the everyday economy, which continues to occupy a central role in providing jobs and services for local demand and local need throughout Portugal. The need to strengthen local governance, in order to improve planning processes and ensure the delivery of quality local services oriented to local needs, requires not only changes to formal political systems, but also an active citizenship and participatory culture that builds upon existing civic values and embraces new ones.

Policy frameworks need to promote and retain a holistic vision of local and regional development that seeks to integrate economic, social and environmental needs whether in the rural interior or the urban neighbour-hood. Such an integrated vision of work, culture and everyday life embedded within specific places has a strong tradition in Portugal, particularly within rural life. In the growing urban spaces, with their new forms of interaction, social division and daily rhythms, the challenge is to adjust to the new urban realities of intensity and difference, but in a manner that can retain the tradition of strong place-based communities.

Notes

1   At the time of writing the full results of the 2001 population census were not available for analysis. However some preliminary results had been published and selected use has been made of these (INE, 2001).

2   The EU operates a system of classification of territorial units (NUTs) which moves from the national level (Level I), to the regional level (Level II) (in Portugal this level corresponds to the areas of the regional planning commissions), and so on to smaller scale territorial units.

3   Since the introduction of the current system of EU regional funding in 1989, all of Portugal has been designated as an 'Objective One' region, that is a region where 'development is lagging behind'. The key criteria for this classification is that GDP per capita within the region is only up to 75% of EU average.

4   In 1999 sales per worker (in US dollars) in the mould industry were 51,145 in Portugal, compared to 101,500 in Germany, and 180,821 in Japan. Labour costs in Portugal were 6.5 Euros per hour for a skilled tool designer, as compared to 18.2 in Germany, and 33.7 in the United States (Cefamol, 2001).

5   Since 1989 Portugal has received European regional funding via a series of 'Community Support Frameworks' (CSFs), which prioritise development objectives in line with EC guidance and set out funding arrangements. The first of these, CSF I ran from 1989-1993, the second, CSF II, 1994-1999, and the third, CSF III, 2000-2006. Before 1989 Portugal also received EU regional funds under a range of other policies (see Syrett 1995).

6   The second CSF received funding of 26,678 million ECU, 52% of which came from the Structural Funds. In addition further EU finance was received via the Cohesion Fund and a range of Community Initiatives.

7   It is estimated that Portuguese farmers receive only 10% of what higher income Danish farmers receive under existing CAP arrangements (Financial Times, 2001).

References

Baklanoff, E.N. (1978), *The Economic Transformation of Spain and Portugal*, Praeger, New York.

Barreto, A. (1996), 'Três Décadas de Mudança Social', in A. Barreto (ed) A *Situação Social em Portugal*, 1960-1995, Instituto de Ciências Sociais, Universidade de Lisboa, Lisbon, pp.35-60.

Caetano, L. (1995), 'Distritos Industriais no Desenvolvimento Recente de Portugal: O Caso do Centro-Litoral', *Cadernos de Geografia*, Instituto de Estudos Geográficos, Universidade de Coimbra, Coimbra, no.14, pp.3-16.

Cefamol (2001), *Situação da Indústria de Moldes em 2000*, Cefamol, Marinha Grande.

Commission of the European Communities (CEC) (1990), *Community Support Framework 1989-93: Portugal*, Commission of the European Communities, Brussels.

Corkill, D. (1999), *The Development of the Portuguese Economy: a Case of Europeanization*, Routledge, London.

Costa, E. and Costa, N. (1996), 'Reflexos Territoriais do Processo de Reestruturação Industrial em Portugal Continental na Década de Oitenta', *Finisterra*, vol.XXXI, no.62, pp.69-92.

Ferrão, J. (1996), 'Três Décadas de Consolidação do Portugal Demográfico Moderno', in A. Barreto (ed), *A Situação Social em Portugal, 1960-1995*, Instituto de Ciências Sociais, Universidade de Lisboa, Lisbon, pp.165-190.

Ferrão, J. and Mendes Baptista, A. (1992), 'Industrialisation and Endogenous Development in Portugal: Problems and Perspectives', in G. Garofoli (ed) *Endogenous Development and Southern Europe*, Avebury, Aldershot, pp.195-212.

Financial Times (2001), *Portugal: Annual Country Report*, October 24.

Gaspar, J. (1997), 'Lisbon: Metropolis Between Centre and Periphery' in C. Jensen-Butler, A. Shachar and J. Weesep (eds) *European Cities in Competition*, Avebury, Aldershot, pp.147-178.

Gaspar, J. and Jensen-Butler, C. (1992), 'Social, Economic and Cultural Transformations in the Portuguese Urban System', *International Journal for Urban and Regional Research*, vol.16, no.3, pp.442-461.

Gaspar, J. and Williams, A.M. (1991), *North and Central Portugal in the 1990s: A European Investment Region*, Economic Intelligence Unit, London.

Instituto Nacional de Estatística (INE) (1991), *XIII Recenseamento Geral da* Populacão, INE, Lisbon.

Instituto Nacional de Estatística (INE) (1993), *Alterações Demográficas nas Regiões Portuguesas entre 1981-91*, INE, Lisbon.

Instituto Nacional de Estatística (INE) (2001), *XIV Recenseamento Geral da Populacão (Resultados Preliminares)*, INE, Lisbon.

Lewis, J.R. and Williams, A.M. (1981), 'Regional Uneven Development on the European Periphery: the Case of Portugal 1950-78', in *Tidjschrift voor Economische en Sociale Geografie*, vol.72, no.2, pp.81-98.

Lopes, J. Silva (1996), 'A Economia Portuguesa Desde 1960' in A. Barreto (ed) *A Situação Social em Portugal*, 1960-1995, Instituto de Ciências Sociais, Universidade de Lisboa, Lisbon, pp.233-364.

Ministério de Plano e Administração do Território (MPAT) (1993), *Desenvolvimento Regional em Portugal*, MPAT, Lisbon.

Minstério do Planeamento (1999), *Portugal: Plano de Desenvolvimento Regional, 2000-2006*, Ministério do Planeamento, Lisbon.

Minstério do Planeamento (2000), *Quadro Comunitário de Apoio III: Portugal 2000-2006*, Ministério do Planeamento, Lisbon.

Observatório do Emprego e Formação Profissional (1995), *Estudo Socio-Económico da Marinha Grande e Área Envolvente*, Observatório do Emprego e Formação Profissional, IEFP, Lisbon.

Portas, N., Domingues, A., and Guimarãis, A.L. (1998), 'Portugal', in L. van den Berg, E. Braun, and J. van der Meer (eds) *National Urban Policies in the European Union*, Ashgate, Aldershot, pp.290-324.

Silva, C.N. (1999), 'Local Government, Ethnicity and Social Exclusion in Portugal', in A. Khakee, P. Somma, and H. Tomas (eds) *Urban Renewal, Ethnicity and Social Exclusion in Europe*, Ashgate, Aldershot, pp.126-147.

Simões Lopes, A. (1983), 'Desenvolvimento Regional: O 'Estado da Arte em Portugal, ou a Política da Ausência de Política', *Estudos de Economia*, vol.III, pp.231-35.

Smallbone, D., Cumbers, A., Syrett, S. and Leigh, R. (1999), 'The Single European Market and SMEs: a Comparison of its Effects in the Food and Clothing Sectors in the UK and Portugal', *Regional Studies*, vol.33, no.1, pp.51-62.

Syrett, S. (1995), *Local Development: Restructuring, Locality and Economic Initiative in Portugal,* Avebury, Aldershot.

Syrett, S. (1996), 'International Regulation and Regional Development: Consequences of the Single European Market for Small and Medium Sized enterprises in Portugal', in *European Planning Studies*, vol.6, no.4, pp.739-756.

Syrett, S. (1997a), 'Cavaco Silva, the European Union and Regional Inequality: Regional Development Policy in Portugal, 1985-95', *International Journal of Iberian Studies*, vol.10, no.2, pp.98-108.

Syrett, S. (1997b), 'The Politics of Partnership: the Role of Social Partners in Local Economic Development in Portugal', *European Urban and Regional Studies*, vol.4, no.2, pp.99-114.

Vale, M. (1998), 'Industrial Restructuring in the Lisbon Metropolitan Area: Towards a New Map of Production?' in T. Unwin (ed) *A European Geography*, Longman, Harlow, pp.178-181.

4 Tourism in Portugal: From Polarisation to New Forms of Economic Integration?

ALLAN M. WILLIAMS

Introduction: a Changing Industry in a Changing Country

Tourism has been one of the most dynamic economic sectors in the second half of the twentieth century, and its role in the restructuring of European economies is widely acknowledged (Williams and Montanari, 1995). This chapter explores the economic contribution of tourism to the Portuguese economy in this period, when the country progressed from relative under-development, and a semi-peripheral economic position (Seers et al, 1979), to founder membership of the European Monetary System and rapid convergence towards average EU economic levels of GDP. Tourism played an important role in this process, but this chapter focuses on in its contribution to the transformation of the Portuguese economic space. Particular consideration is given to the spatial articulation of the economic impact of tourism and whether it has shifted from a high degree of spatial polarisation to a more dispersed development pattern. The Algarve is the region which epitomises this transformation, for in a few decades it has passed from relative tourism obscurity to being one of the major mass tourism destinations in Europe.

While the changing position of the Algarve will be analysed later in the chapter within a broad political economy framework, an insight into the scale and intensity of change is obtained by comparing three travel commentaries on the region. In the mid 1930s, in Playtime in Portugal, John Gibbons (1936, p.132) visited an Algarve where the tourist was still a novelty, and tourism facilities were thin on the ground. He commented that:

> There is everything in Faro for every taste, except an hotel... There is one quite good one where English tourists stop at Praia da Rocha, and there are at least two more at Monchique Caldas; also at Vila Real there is Hotel Guadiana kept by a German with a perfect knowledge of English. But that is all.

Even as late as 1961, the Fodor Guide (1961, p.314 & p.355) still found a largely 'undiscovered Algarve' that had yet to become a significant object of the 'tourist gaze' (Urry, 1990). Then, at the dawn of the age of mass tourism, Lisbon was the focus of tourism and the Algarve was considered relatively isolated.

> Portugal, herself responsible for the discovery of so much of the surface of the earth, is still a country to be discovered... The problem of touring Portugal is simplified by the fact that wherever you want to go, you start from Lisbon. Planes and ships set you down there, all roads and railways entering the country are headed for Lisbon. There are not half a dozen different ways of exploring Portugal. There is only one.

Some three decades later, the Rough Guide (1994, p.351) found no traces of this 'undiscovered' Algarve. Instead, the tourism experiences of the Algarve had been commodified as a product of mass consumption, and its territorial structures were dominated by tourism:

> The Algarve has attracted more tourist development than the rest of the country put together. In parts this has all but destroyed the charms that it was intended to exploit. The strip of coat west from Faro to Lagos has suffered most, with its endless villa complexes creating a rather depressing style Mediterranean suburbia.

It would be misleading to suggest that Portuguese tourism became synonymous with the Algarve, even in the mainland of Portugal (that is, excluding Madeira and the Açores). But the growth of mass tourism in the Algarve, alongside the continuing expansion of tourism in Lisbon, meant that the country developed along spatially bi-polar lines, whilst Madeira was the only other region with a significant tourism presence. The key question is whether a more diversified and spatially dispersed pattern of tourism has begun to evolve at the end of the twentieth century, and whether this will be extended in future. In other words, what has been and will be the tourism destiny of Portugal.

A New International Division of Tourism Labour: Portugal as 'Pleasure Periphery'

The growth of Portuguese tourism is most vividly illustrated by the international sector. Whereas, in the 1950s, Portugal attracted less than 1 million visitors, by 1997 this had increased to some 23 million, of whom 10 million were tourists (see Figure 4.1). In this instance, tourists are defined as overnight visitors. Although a chaotic conceptualisation which confuses business tourism, holidays for pleasures and visits to friends and families, it reflects the availability of secondary statistics, and so is utilised in this chapter. The general conditions favouring the growth of international tourism and, in particular, the Mediterranean region becoming the 'pleasure periphery' (Turner and Ash, 1975) of Europe are discussed elsewhere (Montanari, 1995; Shaw and Williams, 1994; Williams, 1997), and here we give only a brief account. From the 1950s, but more especially the 1960s, there was a gradual growth in international tourism to the Mediterranean which was fuelled by the virtuous circle of the social construction of mass foreign tourism around sea and sunshine holidays, falling transport costs (monetary and otherwise), and the economies of scale and marketing power of the major tour companies, which exerted a strong downwards pressure on real price levels.

Portugal, although it had a long history of international tourism (Pina, 1988), based on the particular attractions of Madeira, Lisbon and the numerous spas, came relatively late to mass tourism. To some extent this reflected the 'undiscovered' nature of the Algarve, which tended to be excluded from the social construction of the Mediterranean (see the definitions in Montanari, 1995) as a tourist destination: a situation exemplified by the absence of the region from guide books on the Mediterranean. The Portuguese government was also initially less proactive than its Spanish neighbour in promoting

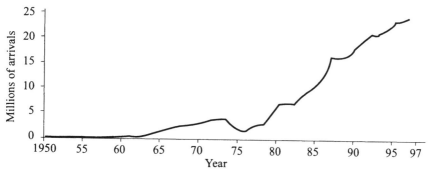

Source: INE, 1998

Figure 4.1 International Visitor Arrivals in Portugal, 1950-97

international tourism. However, after Faro airport had been inaugurated in the mid-1960s, opening up the Algarve to international tourism and especially to charter flights, the region became embedded in the mental maps of northern European tourists and, perhaps more importantly, in the programmes of international tour companies. Madeira, because of its remoter location and lack of beaches, played a lesser role in this initial boom in mass tourism, as did the colder water resorts of the Atlantic coast of the mainland.

The expansion of mass tourism gathered pace in the 1960s, but was disrupted by the political transition following the overthrow of the Salazar/ Caetano regime in 1974. Portugal lost some four years of growth in this period, before expansion was renewed with increasing vigour in the 1980s and the 1990s (Lewis and Williams, 1998). The phenomenal growth is reflected in the numbers of international visitors (tourists and day excursionists) to Portugal, which were less than 100,000 in 1950 but had risen to 23 million by 1997, and to an estimated 26 million by 1998 (Direcção Geral do Turismo, 1998). Tourist numbers also increased to 10.1 million by 1997, and to an estimated 11.2 million in 1998, so that these were broadly equal to the resident population of the country; a ratio which was beginning to approach the tourist-population balance recorded in Spain. While this increase partly reflected the intensification of business travel consequent upon Portugal's economic expansion after 1986, it was driven above all by holiday tourism, and especially by beach tourism in the Algarve.

In common with most Mediterranean destinations, Portugal's international market was dominated by a relatively few source countries. This reflected the realities of the distribution of consumer expenditure in northern Europe, but also the particular strength of international tour companies in Germany and the UK (and, to a lesser extent, the Benelux countries and Scandinavia). The degree of market concentration depends on which particular set of statistics is used to calculate it. Statistics on excursionism, or the total number of arrivals, tend to give Spain a strong position in the tourism market. However, data on overnights are used here as providing a more accurate reflection of potential economic impacts. These statistics reveal that, even in 1960, Portugal was highly dependent on a small number of countries of tourist origin (Table 4.1). The UK accounted for more than a fifth of all tourists at that date, and France for almost another one fifth. Over time, Portugal has become even more dependent on the British market which accounted for 29.3 per cent of all overnights in 1997. There has been a rank reordering of the other national market segments; Germany had clearly emerged in second position, with almost a quarter of overnights in 1997.

These gains were matched by a decline in the French share of the market, partly reflecting the lesser importance of tour companies in that market segment.

The expansion of international tourism resulted, therefore, in Portuguese dependence on both mass summer tourism and a relatively small number of destinations: more than one half of all overnights in 1997 were accounted for by two national market segments. Not surprisingly, there have been increasing criticisms (Smith and Jenner, 1998) of both the level of dependency and the increasingly mass nature of the tourism product which, initially, state policy had sought to locate at the upper end of the mass tourism product range (Pina, 1988; Cavaco, 1979). There were real concerns, expressed in both the press and policy reviews, about how the Algarve in particular had gone 'down-market' in the 1980s and 1990s. Comparisons were frequently made to the Spanish 'costas', which were precisely the models that Portugal had sought to avoid.

In response to these trends in mass tourism, the *Plano Nacional de Turismo*, 1986-89 (Secretaria de Estado do Turismo, 1986), as early as the mid-1980s, had made product diversification one of the overarching goals of Portuguese tourism. Seeking to commodify the attractions of rural Portugal, national and local state policies have supported the sector via marketing activities and capital subsides to improve the accommodation stock (Moreira, 1994). The result has been the creation of a significant stock of accommodation, especially in northern Portugal (Table 4.2), which is not

Table 4.1 Tourist Origin by National Market Segmentation, 1960 and 1997

	Percentage of Total International Visitor Nights in Licensed Accommodation	
	1960	1997
Germany	6.9	24.0
Spain	8.8	7.5
UK	22.3	29.3
France	17.6	4.2
USA	4.8	2.6

Source: Direcção Geral do Turismo,1986; INE, 1998

only a counterbalance to the concentration of mass tourism in the Algarve, but has also diffused tourism into some of the more rural and isolated areas in the country. However, the scale of this initiative remains limited. Even in 1997, there were only some 5,000 rural tourism bed spaces in the whole of Portugal, while the Algarve had almost 90,000 beds. The aggregate economic impact of rural tourism has been even less, given relatively low occupancy rates; only 11 per cent in the sector compared to the national average of 45 per cent.

The seemingly inexorable slide in the market position of Portugal has not been due to a lack of investment. Over time there has been massive investment in tourism accommodation which has been significantly upgraded. For example, between 1980 and 1994, the stock of hotel accommodation increased by 52 per cent while that in the more basic *pensões* decreased by 4 per cent. However, the expansion of the accommodation stock has brought its own problems; the maintenance of relatively high occupancy rates has been dependent on international markets, and especially on the extension of package holidays to successively lower income strata in northern Europe. In addition, private sector investment in accommodation and other facilities has not been matched by similar state investment in basic tourist and social infrastructures. At times, tourism development in the

Table 4.2 Regional Distribution (%) of Rural Tourism* Provision in Mainland Portugal, 1997

Costa Verde	32
Costa de Prata	12
Costa de Lisboa	6
Montanhas	26
Planicies	20
Algarve	4
Total	100

* includes rural tourism, agritourism and *turismo de habitação*

Source: Direcção Geral do Turismo, 1998

Algarve - and increasingly in Madeira - has resembled 'private affluence amongst public squalor', with luxurious hotels being surrounded by poorly maintained landscapes, chaotic transport, and various forms of pollution. Portugal remains very much part of the (relatively) low cost pleasure periphery of northern Europe.

In contrast to the phenomenal growth of international tourism, domestic tourism expansion has been relatively muted. For much of the twentieth century, holidays away from home remained the almost exclusive preserve of the urban middle classes so that, given the narrowness of this social strata, domestic demand was limited (Cavaco, 1979). Instead, for most of the population leisure away from home was limited to well established patterns of daily excursionism from the main cities to nearby coastal areas, and reliance on accommodation provided by friends and family, often in rural areas (Cavaco, 1995); the latter reflecting large scale rural-urban migration in the twentieth century. Therefore, Portugal had one of the lowest rates of domestic tourism participation in Europe, as a consequence of its relatively limited urbanisation and polarised, and overall low, income levels (Shaw and Williams, 1998). Since the mid-1980s, sustained economic recovery has fuelled a consumer boom which has incorporated increased tourism spending. Whereas 5 million holidays were taken by the Portuguese in 1986, by 1994 this number had risen to 7.3 million. Yet, according to national surveys, only 42 per cent of the Portuguese still spent holidays away from home, even in 1998 (Direcção Geral do Turismo, 1999). Not surprisingly, only some 29 per cent of all overnights are provided by the national market, which is one of the lowest proportions in Europe.

Against this background of significant dependence on small numbers of international market segments, the role of tourism in the Portuguese economy is now considered.

The Economic Role of Tourism: a Spatially Polarised Economic 'Blessing'?

The contribution of tourism to the Portuguese national economy has long been recognised. For example, as early as the 1970s, Baklanoff (1978) emphasised the significant contribution of tourism (and emigrant remittances) to the current account. Not only did tourism foreign exchange earnings serve to cover part of the merchandise trade gap, but they also helped to finance imports of capital and intermediate goods which were critical to the industrial transformation in the 1960s and early 1970s. Over time, while the absolute surpluses generated by tourism have continued to grow, their relative importance has declined due to the increasing sophistication of the economy. This is particularly true of the period of sustained growth after 1985. In that year, tourism net earnings covered 80 per cent of the current account deficit. By 1997 there was a net balance on the tourism account of $2.2 billion, which was equivalent only to approximately 40 per cent of the current account deficit.

There were, however, other more worrying reasons for the declining importance of tourism earnings. Most importantly, there has been a long term decline in real spending by tourists; 34 per cent on a per capita basis between 1980 and 1996 (Smith and Jenner, 1998). One reason for this is the growth in tourism from Spain, especially after their joint entry to the EU in 1986. Spanish tourists tend to come on shorter visits, averaging three days, compared to Northern Europeans, averaging 11-15 days. However, there are also other reasons for the decline in real income, and these are related to the changing nature of mass tourism. Data on per diem expenditure, disaggregated by nationality, and reliance on organised tourism, provide insights into this problem (Table 4.3). If France is excluded from the analysis, on the basis that many of the tourists are first or second generation Portuguese emigrants whose expenditure represent various forms of familial trans-border financial transfers, then a clear pattern emerges. Tourists from northern Europe are

Table 4.3 Tourism Spending and Reliance on Organised Travel: National Market Segmentation, 1997

	Expenditure per diem (1.00=national average)	% using travel agencies
Germany	0.73	76
Spain	1.62	34
France	3.35	51
UK	0.70	87
USA	2.32	71
Total	1.00	69

Source: INE, 1998

comparatively low spending and relatively dependent on travel agencies (used here as a surrogate for tour companies, for which data are not available). In contrast, tourists from Spain tend to organise their own holidays, and are relatively high spending; even though the latter is partly related to an element of shopping tourism amongst Spanish visitors.

There are also critical differences between individual and 'package' tour tourism more generally. In the organised tourism sector, there is severe downward pressure on prices via the major tour companies considerable oligopsonistic powers over suppliers in a highly competitive market, where the tourism product is little differentiated (Williams, 1995). In contrast, the American market segment, which is relatively more skewed to higher income groups, is characterised by more than three time higher per diem spending. Given the powers of the tour companies, as the primary inter-mediaries between northern European tourists and Portuguese suppliers of tourism services, and given the essentially 'placeless' nature of package Mediterranean beach holidays, it is difficult for hotels in the Algarve to resist the inexorable pressures to reduce their real prices (and income).

As a result of the importance of the international market for Portuguese tourism, changes in tourism account earnings are closely reflected in the data on production and employment. These are discussed in some detail elsewhere (Lewis and Williams, 1998; Cunha, 1986). Here, the focus is on the spatial impact of tourism and addressing two main questions. First, to what extent are the economic impacts of tourism spatially polarised, that is to say regionally uneven. Second, how does this contribute to the overall pattern of regionally uneven development in Portugal: is tourism a force for convergence or divergence? The overall distribution of tourist activity is summarised in Table 4.4 and Figure 4.2, which provide estimates of the distribution of tourists and tourism expenditures, although these are necessarily approached with caution given the unreliability of many sub-national data series.

The regional distribution of tourism, as would be expected, is highly polarised. At the tourism region level, three regions alone accounted for more than 90 per cent of all recorded overnights in 1997: the Algarve, Lisbon and Madeira. The first of these accounted for more than one half of the total, a level of regional polarisation which is almost unsurpassed in Europe (see Shaw and Williams, 1998). Second place was taken by the Costa de Lisboa (a combination of Lisbon and adjoining coastal resorts such as Estoril), and third place by Madeira. Elsewhere in the country, the Costa Verde, which includes the business tourism centred on Porto, accounted for a further nine per cent of overnights. There are also differences in terms of market segments. Foreign tourism has a far more polarised distribution than domestic tourism: this is due to the higher level of business tourism in the latter, and the legacy of intra-regional holiday tourism trips (for example, from the northern and central interior to the adjacent Atlantic coastal area) (Lewis and Williams, 1998).

The distribution of receipts, while also uneven, is less extremely polarised. This is mainly due to the relatively low receipts in the Algarve, which can be seen as the economic outcome of the particular evolution of the mass tourism model: one dominated by foreign based and owned tour companies. Not only do these secure low prices from local establishments, and retain a high proportion of the total price of the package holiday in the country of origin, but they have also tended to channel increasingly lower income and lower spending tourists to the Algarve. In contrast, the Lisbon region, which has less than a quarter of overnights has one third of tourism receipts, reflecting its significant international and national business tourism sectors, as well as higher spending urban cultural tourism. Elsewhere in the mainland, all the other regions secure higher shares of receipts than of overnights, but, these are not especially significant. Madeira has a slightly higher share of receipts than of overnights; although its tourism industry is increasingly subject to similar pressures to those experienced in the Algarve, it still continues to attract relatively higher spending tourists.

Table 4.4 The Regional Distribution of Tourism, 1997

			Employment	
Tourism region	% overnights	% receipts	Absolute	%
Costa Verde	9.1	11.4	4,149	10.8
Costa de Prata	6.9	7.2	3,168	8.3
Costa de Lisboa	23.8	33.3	8,837	23.0
Montanhas	3.4	4.2	2,066	5.4
Planícies	3.1	3.9	1,723	4.5
Algarve	53.1	39.8	12,117	31.6
Açores	1.7	2.4	984	2.6
Madeira	17.0	18.8	5,344	13.9
Total	100.0	100.0	8,388	100.0

Source: INE, 1998

Figure 4.2 Major Tourism Regions

Employment has an even less polarised distribution, with the Algarve and the Costa de Lisboa accounting for more than one half of all jobs, and Madeira for just over 13 per cent. Given that these three areas have almost all the five star hotels in Portugal, and given also that employment-guest ratios increase sharply with hotel quality (Lewis and Williams, 1998), this at first seems rather surprising. The answer, however, lies in the nature of the employment, for the main destinations have higher levels of full time waged positions. In contrast, there are relatively more part time jobs, or positions filled by non-remunerated family in the other regions: for example, the latter only account for 1.9 per cent of jobs in the Algarve compared to 4.2 per cent nationally. Moreover, there are also more family employees in the generally smaller establishments outside of the Algarve and Lisboa, and it is likely that, in reality, many of these may be substantially under-employed in their tourism jobs.

The overall conclusion is that tourism clearly does have a highly polarised distribution and contributes to uneven regional development in Portugal. Whether it exacerbates or modifies the overall regional pattern is a more complex issue. With respect to business tourism, this simply reflects underlying regional economic structures. However, most forms of pleasure tourism (excluding the urban cultural) tend to be located in peripheral regions, because it is precisely the attractions of 'unspoilt' nature and coastal areas which attract tourists seeking rest and recreation, and a contrast to the intensification of working and living conditions in metropolitan areas. This can have an enormous economic impact on individual communities in rural areas, but relatively minor implications for the economic structures and overall trajectories of these regions. However, there are two exceptions to this: the Algarve and Madeira, where tourism is the leading economic sector. In the former, tourism has made a major contribution to the region's overall economic improvement relative to other regions, and its convergence towards the national mean on many economic indicators (Gaspar, 1993).

Reflections on Portuguese Tourism: Globalisation, Polarisation and New Forms of Mobility

While reference has so far been made to the European context of Portuguese tourism, the concept of globalisation provides a basis for a fuller appreciation of its structure and future trajectory. Globalisation is, of course, a contested concept (Amin and Thrift, 1994), but here we adopt Held's approach, which is centred on the interactions between the global and the local. He argues that globalisation represents:

> ...the stretching and deepening of social relations and institutions across space and time such that, on the one hand, day-to-day activities are increasingly influenced by events happening on the other side of the globe and, on the other, the practices and decisions of local groups or communities can have significant global reverberations (Held, 1995, p.20).

Two important conclusions can be drawn from this literature for our understanding of Portuguese tourism. First, globalisation is essentially about the intensification of competition and the trajectory of tourism in Portugal has to be understood in terms of global as well as European competition. Given that Portuguese tourism is constituted of a number of different tourism products, each of which has a distinctive territorial structure, it is very much the case that the trajectory of tourism development in each region and locality has to be understood in this way. Thus Madeira, the Algarve, agritourism in the Alentejo, and urban cultural tourism in Lisbon, all face very different forms of competition, but with the commonality that this is increasingly globalised. Secondly, localities and regions are not passive recipients of globally originating tourism processes. Local groups can have 'significant global reverberations', and they can organise in order to 'contest the global'. Thus local areas can develop effective strategies to re-position themselves in terms of global tourism, or at least to mediate the effects of global competition.

A second important argument to note is that it tourism needs to be understood in context of overall systems of mobility, and not as an isolated form (Williams and Hall, 2000). A systems perspective provides an useful starting point here:

> In the context of an increasingly inter-connected world, international population movements can naturally be seen as complements to other flows and exchanges taking place between countries. Indeed international migrations do not occur randomly but take place usually between countries that have close historical, cultural or economic ties (Kritz and Zlotnik, 1992, p.1).

Tourism has to be seen as a component in larger cycles, or reciprocal flows, of population movements. Examples of this have already been encountered in the preceding discussion. First, in respect of high levels of expenditure by tourists originating in France, a fact that is related to the history of Portuguese mass emigration to that country, and the subsequent generation of return flows to visit families and friends. Secondly, there are relatively strong urban-rural domestic tourism flows in Portugal, which are rooted in an earlier generation of rural-urban migration movements. Over time, both these flows are likely to become moderated, as the links of succeeding generations with the areas of origin are seen to weaken. There are, however, likely to be new forms of tourism flows generated by, and generating, other forms of mobility, as is seen below in respect to international retirement.

Within this general theoretical context, we can explore the reasons for the high degree of polarisation in the Portuguese tourism industry.

In particular, there are three main reasons for the existing pattern of spatial concentration, apart from obvious regional differences in the distribution of natural and cultural attractions. These are structural rigidities arising from the inter-relationship between fixed capital and tourism flows, the effects of competition on mass tourism, and the complementarities of different forms of population flows.

Mass tourism requires substantial infrastructural investments, particularly social investments in transport and environmental services, and private sector investment in accommodation and catering services. There are two consequences of this. First, the scale of the investment places an imperative on the destination region or country to generate and maintain tourism at sufficiently high levels to recoup the original costs of the investments, and the further investments required to renew these. This is clearly the case with both Madeira and the Algarve, where there have been massive investments in fixed capital, especially the airports, roads, and hotels/apartments. Secondly, the generation of mass tourism flows implies a high degree of path dependency in future tourism flows. While some mass tourists do seek 'the challenge of the new', and there are also cycles of fashion amongst tourism destinations, the initial waves of tourism create a legacy which shapes later flows. Many tourists develop attachments to particular places, and make repeat visits, while their recommendations to other tourists often play a critical role in determining destination choice (Nolan, 1976).

Polarisation is also reinforced by the very nature of global competition. As a largely undifferentiated product, competition in respect of Mediterranean beach tourism largely centres on price issues. This has an important bearing on the behaviour of tour operators. While it is in the interests of tour companies to diversify the range of destinations that they offer so as to minimise risk, there are also economies of scale (related to the economics of filling whole aircraft, marketing costs, and maintaining representatives in the destination resorts) and market dominance reasons for channelling demand to a limited number of destinations. To this should be added the fact that there is a very high degree of concentration in the tour operator sector in Portugal's two main markets, Germany and the UK. In both of these countries more than one half of the total market for package holidays is accounted for by only three companies. Amongst UK companies, Thomson has a particularly large share of the market for the Algarve (Monopolies and Mergers Commission, 1989). As a result, there is a particularly favourable combination of national market segmentation, tour company concentration, and the economics of mass tourism, which favour spatial polarisation. While the same conditions do not apply in respect of other tourism products, the Algarve and Madeira alone account for more than two thirds of tourist overnights in Portugal.

The third reason for the polarisation of tourism centres on the complementarities of international population flows. There are two main expressions of this. First, there is the increasingly multi-functional nature of tourism trips, leading to what Urry (1990) has argued may be termed post-modernist tourism. This is particularly evident in the way that business trips are often combined with aspects of leisure tourism. For example, business tourists to Lisbon are taken to expensive restaurants and perhaps to the theatre by their hosts. They may also extend their visits by an extra day or two for some leisure activities, perhaps taking members of their families with them for this purpose. The important point here is that the uneven economic development which produces a concentration of business tourism, also leads indirectly to reinforcing cultural and pleasure tourism visits to places such as Lisbon and Porto.

Retirement migration provides a second example of the way in which international tourism complements other population flows in an increasingly inter-connected world (Kritz and Zlotnik, 1992). International retirement migration from northern to southern Europe is a relatively recent phenomenon, but an increasingly well-documented one (King et al, 1998; Williams et al, 1997; Williams and Patterson, 1998). The Algarve is one of the principal destinations for such population flows, and according to the best available estimates there are at least 10,000 retired or semi-retired British people living in the region. Although the largest national expatriate group, there are also increasing numbers of Germans as well as smaller communities from most other northern European countries. There is evidence that previous tourist visits to the Algarve provided the major previous connection that most retirees had with the area, and which brought it within their retirement search spaces. In this instance, tourism has generated retirement flows, ranging from permanent, to various forms of temporary migration and long term tourist stays. As has been noted, elsewhere, such inflows can help to sustain the economic bases of tourist destinations as they enter the later stages of the tourism product life cycle. Moreover, the retirees also become the focus of new rounds of tourist visits by family and friends (Williams et al, 2000).

While, thus far, we have concentrated on polarisation, the earlier quotation on globalisation also provides a framework for understanding the counter-balancing effects of local economic development strategies. In the face of increasing pressures to restructure local economies, the local state has become increasingly entrepreneurial, and, often in partnership, has taken the lead in promoting local economic development strategies. Tourism, because of the buoyancy of demand in the sector, the ubiquity of potential tourism attractions, and relatively low capital start up costs (Williams and Shaw, 1998), has often provided the basis for such strategies. This is evident, for example, in the strategies to develop tourism in the Baixo Mondego and in the Alentejo. But perhaps the most spectacular example was the 1998 Lisbon EXPO which used the tourism income generated by a mega event as a catalyst for a major redevelopment programme for an economically depressed and environmentally damaged area on the city's waterfront. With the exception of Lisbon, the overall effect of these initiatives has been to foster a more dispersed pattern of tourism activity. This has considerable implications for individual areas, but it has to be seen against the strong processes of tourism polarisation in Portugal.

The Tourism Future of Portugal: Ever More Polarised?

There are widespread predictions that tourism will continue to expand globally, and that Portugal will have a share in this. Here we consider whether the high level of polarisation is likely to continue, or will be modified by new forms of tourism investment and of tourist preferences and behaviour. Any such changes, however, are likely to be incremental for there is a high degree of path dependency inherent in future tourism development, given both accumulated fixed investment, and the way in which particular Portuguese tourism products, especially the Algarve, have evolved strong place images, and prominent positions in most tourists' perceptions of Portugal as a holiday destination.

Turning first to national tourism, this is a sector where strong growth can be expected, in view of the current low rates of participation. There are both economic and cultural reasons for this, but they are likely to weaken as Portugal continues to converge on average EU GDP and income per capita levels. The question then is whether the increased leisure time and improved standards of living of the Portuguese will lead to increased tourism and, if so, whether they will prefer domestic or foreign destinations. At present, foreign holidays are not a priority for the Portuguese: only 17 per cent went abroad for holidays in 1998 (Direcção Geral do Turismo, 1999). This is exceptionally low in comparison to most other European countries. The proportion is likely to rise in future, and the strongest indication of this is to be found in the holiday behaviour of different age groups. Amongst those who took holidays away from home in 1998, only 31 per cent of those aged over 65 went abroad, compared to 74 per cent of those younger than 34. In addition, there is also evidence of a long term trend for the outflow of tourism expenditures from Portugal to grow more rapidly than the inflow. Moreover, the effects of globalisation in terms of opening up competition from increasing numbers of potential destinations around the globe, is also likely to lead to increasing numbers of holidays been taken outside of Portugal. Despite the anticipated relative shift to foreign holidays, there is also likely to be continued strong absolute growth in domestic tourism. Existing trends suggest that this is likely to contribute to further tourism polarisation. Between 1992 and 1997, while there was a 65 per cent increase in domestic flows to beach resorts, there was a 24 per cent decrease in those to rural destinations (Smith and Jenner, 1998). Increasing numbers of Portuguese tourists are likely to favour the Algarve, given its climatic advantages, and accumulated fixed tourism investment, thereby reinforcing polarisation. Elsewhere, these beach-orientated flows will add to the existing pattern of intra-regional polarisation.

Given the overall importance of inbound international tourism, changes in this sector are likely to be more significant for the immediate trajectory of Portuguese tourism. The World Tourism Organisation predicts that tourist numbers will increase from 613 million in 1997 to 1.6 billion by 2020. Overall, the global shares of both Europe and the Mediterranean are likely to shrink, but strong absolute growth is expected to continue (Jenner and Smith, 1993; Montanari, 1995). There are signs that the nature of demand is changing, and that there will be a continuing shift to more individualised and flexible patterns of holiday taking, with more but shorter holidays replacing one traditional long holiday. There may, therefore, be relatively higher growth rates in the cultural, rural and other niche tourism sectors, compared to mass tourism. If this is so, there may be a reduction in the degree of relative polarisation, and regions other than the leading three may be able to secure a larger share of the total tourism market. Against this, there is also evidence that the Mediterranean tourism product is being adapted to changing consumer demands, including those for greater flexibility and individualisation (Marchena Gómez and Rebollo, 1995), and may be able to maintain its market share.

Finally, given that the peak growth rates in Portuguese tourism did not occur until relatively late compared to other Mediterranean destinations (Williams, 1997), further expansion of international retirement migration can be expected. This will also be favoured by the continuing ageing of the European population, and the polarisation of income so that there are increasing numbers of retired and early retired persons with the means to relocate partially or totally to the south of Europe (Williams et al, 1997). One indicator of the potentially strong growth is that, between 1994 and 1997 alone, there was a 19 per cent increase in the number of British pensions paid in Portugal. Although data is not available on their regional distribution, it can be presumed that a large proportion are destined for the Algarve, which is the dominant retirement destination for northern Europeans (Williams and Patterson, 1998). This is one reason behind the growth of legally-registered foreign-owned residences in the Algarve; from 1,853 in 1975 to 23,122 in 1998 (INE, 1998). In the same period, the Algarve share of all such property in Portugal increased from 6 per cent to 13 per cent. Given the current distribution of foreign tourism within Portugal, retirement migration is likely to continue to be concentrated in the Algarve. In contrast, travel costs remain a barrier to in-migration to Madeira.

There are, however, signs that the intra-regional distribution of international retirement in the Algarve is changing (Table 4.5). Almost two thirds of the arrivals from the UK in the 1960s settled in coastal urban areas, either in existing settlements such as Albuferia, or new ones such as Vilamoura. In

Table 4.5 The Changing Geographical Distribution of Retired British Residents in the Algarve, 1961-96

	% living in settlement type:			
	Coastal urbanisation	Coastal rural	Interior & west coast	Total
Date of arrival				
1961-75	64.3	17.9	17.9	28
1976-85	47.6	33.3	19.0	42
1986-90	46.8	29.9	23.4	77
1991-96	36.1	32.8	31.2	61
Total sample	46.2	29.8	24.0	208

Source: Williams and Patterson, 1998, p.149

contrast, only just over a third did so in the 1990s in consequence of a gradual decentralisation into, first, the coastal rural zones, and later the interior and the relatively remote western coast. This was influenced by both push (congestion and other costs associated with continued tourism development) and pull factors (seeking out idealised rural idylls), as well as the general facilitation provided by increased state investment throughout the region. Therefore, while retirement migration is likely to continue, its impact will be different to that in earlier rounds, not least because the Algarve itself has changed.

This last point leads to a concluding comment. While past tourism trends do provide some indication of the likely future trajectory of Portuguese tourism, the precise nature of this will be different, as will its impact. Path dependency does not imply the mechanistic repetition of previous tendencies and structures. Rather, the continuing trends in mass tourism will intersect with many other new and old economic and social processes, including those such as retirement migration which were initially facilitated by tourism. Moreover, Portugal is becoming an increasingly sophisticated and internationalised economy and the impacts of tourism will be both shaped and mediated by this.

References

Amin, A. and Thrift, N. (1994), *Globalization, Institutions and Regional Development in Europe*, Oxford University Press, Oxford.

Baklanoff, E.N. (1978), *The Economic Transformation of Spain and Portugal*, Praeger, New York.

Cavaco, C. (1979), *O Turismo em Portugal: Aspectos Evolutivos e Espaciais*, University of Lisbon, Estudos de Geografia Humana e Regional, Lisbon.

Cavaco, C. (1995), 'Rural Tourism: the Creation of New Tourist Spaces', in A.Montanari and A.M.Williams (eds), *European Tourism: Regions, Spaces and Restructuring*, Wiley, Chichester, pp.127-150.

Cunha, L. (1986), 'Turismo,' in M. Silva (ed), *Portugal Contemporâneo: Problemas e Perspectivas*, Instituto Nacional de Administração, Lisbon, pp.293-312.

Direcção Geral do Turismo (1986), *O Turismo em 1983/84*, Direcção Geral do Turismo, Gabinete de Estatísticas e Inquéritos, Lisbon.

Direcção Geral do Turismo (1998), *Análise de Conjuntura, Boletim 30*, Direcção Geral do Turismo, Lisbon.

Direcção Geral do Turismo (1999), *Férias dos Portugueses 1998*, Direcção Geral do Turismo, Lisbon.

Fodor (1961), *Portugal*, Fodor, London.

Gaspar, J. (1993), *The Regions of Portugal*, Ministry of Planning and Administration of the Territory, Lisbon.

Gibbons, J. (1936), *Playtime in Portugal: An Unconventional Guide to the Algarve*, Methuen, London.

Held, D. (1995), *Democracy and the Global Order: From the Modern State to Cosmopolitan Governance*, Polity Press, Cambridge.

INE (1998), *Estatísticas Demográficas*, Instituto Nacional de Estatísticas, Lisbon.

Jenner, P. and Smith, C. (1993), *Tourism in the Mediterranean*, Economist Intelligence Unit, London.

King, R., Warnes, A.M. and Williams, A. (1998), 'International Migration in Europe', *International Journal of Population Geography*, vol.4, no.2, pp.91-111.

Kritz, M.M. and Zlotnik, H. (1992), 'Global Interactions: Migration Systems, Processes and Policies', in M.M. Kritz, L.L. Lim and H. Zlotnik (eds) *International Migration Systems*, Clarendon Press, Oxford, pp.1-16.

Lewis, J.R. and Williams, A.M. (1998), 'Portugal: Market Segmentation and Economic Development', in A.M. Williams and G. Shaw (eds), *Tourism and Economic Development: European Experiences*, Wiley, Chichester, pp.125-150.

Marchena Gómez, M.J. and Rebollo, F. V. (1995), 'Coastal Areas: Processes, Typologies and Prospects', in A. Montanari and A.M. Williams (eds), *European Tourism: Regions, Spaces and Restructuring*, Wiley, Chichester, pp.111-26.

Monopolies and Mergers Commission (1989), *Thomson Travel Group and Horizon Travel Ltd*, HMSO, London.

Montanari, A. (1995), 'The Mediterranean Region', in A. Montanari and A. M. Williams. (eds), *European Tourism: Regions, Spaces and Restructuring*, Wiley, Chichester, pp.41-65.

Moreira, F. J. (1994), *O Turismo em Espaço Rural*, Centro de Estudos Geográficos, Lisbon.

Nolan. S.D. (1976), 'Tourists' Use and Evaluation of Travel Information Sources: Summary and Conclusions', *Journal of Travel Research*, vol.14, pp.6-8.

Pina, P. (1988), *Portugal: O Turismo no Século XX*, Lucidus, Lisbon.

Rough Guide (1994), *Portugal: the Rough Guide*, Penguin Publications, London.

Secretaria de Estado do Turismo (1986), *Plano Nacional de Turismo 1986-1989*, Grupo Coordenador de Plano Nacional de Turismo, Lisbon.

Seers, D., Schaffer, B. and Kiljunen, M.L. (eds) (1979), *Underdeveloped Europe: Case Studies in Core-Periphery Relations*, Harvester Press, Brighton.

Shaw, G. and Williams, A.M. (1994), *Critical Issues in Tourism: a Geographical Perspective*, Blackwell, Oxford.

Shaw, G. and Williams, A.M. (1998), 'Western European Tourism in Perspective', in A.M. Williams and G. Shaw (eds), *Tourism and Economic Development: European Experiences*, Wiley, Chichester, pp.17-42.

Smith, C. and Jenner, P. (1998), *Portugal, International Tourism Reports*, 1998 volume, no,1, pp.47-76.

Turner, L. and Ash, J. (1975), *The Golden Hordes: International Tourism and the Pleasure Periphery*, Constable, London.

Urry, J. (1990), *The Tourist Gaze*, Sage, London.

Williams, A.M. (1995), 'Capital and the Transnationalisation of Tourism', in A.Montanari and A.M. Williams (eds), *European Tourism: Regions, Spaces and Restructuring*, Wiley, Chichester, pp.163-176.

Williams, A.M. (1997), 'Tourism and Uneven Development in the Mediterranean', in R. King, L. Proudfoot, and B. Smith (eds), *The Mediterranean: Environment and Society*, Edward Arnold, London, pp.208-26.

Williams, A.M. and Hall, M. (2000), 'Tourism and Migration: New Relationships Between Production and Consumption', *Tourism Geographies: International Journal of Space, Place and the Environment*, vol.2.

Williams, A.M., King, R. and Warnes, A. (1997), 'A Place in the Sun: International Retirement Migration from Northern to Southern Europe', *European and Regional Studies*, vol.4. pp.115-34.

Williams, A.M., King, R., Warnes. A. and Patterson, G. (2000), Tourism and International Retirement Migration: New Forms of an Old Relationship in Southern Europe, *Tourism Geographies: International Journal of Space, Place and the Environment*, vol.2.

Williams, A.M. and Montanari, A. (1995), 'Introduction', in A. Montanari and A. M. Williams (eds) *European Tourism: Regions, Spaces and Restructuring*, Wiley, Chichester, pp.1-16.

Williams, A.M. and Patterson, G. (1998), ' "An Empire Lost but a Province Gained": a cohort analysis of British international retirement in the Algarve', *International Journal of Population Geography*, vol.4, no.2, pp.145-56.

Williams, A.M. and Shaw, G. (1998), 'Tourism Policies in a Changing Economic Environment', in A.M. Williams and G. Shaw (eds), *Tourism and Economic Development: European Experiences*, Wiley, Chichester, pp.375-392.

5 International Population Mobility, Immigration and Labour Market Change in Portugal

MARTIN EATON

Introduction

Population mobility has been a fundamental component of Portuguese society for hundreds of years. Indeed, the demographic structure of the country has been intrinsically shaped by the international movements of both the country's domestic and foreign migrant populations (Peixoto, 1996; Fonseca and Cavaco, 1997). In the last 25 years, Portugal has moved from a country associated with large-scale emigration to one typified by a new phase of immigration. While the net migratory balance fluctuated in the 1990s, movements of people into Portugal have had profound spatial impacts. In one of its simplest manifestations, a dual labour market has now emerged covering the official and informal sectors of economic activity, and both are increasingly being fuelled by the employment of foreign immigrant workers. This has led to the repositioning of Portugal within the international division of labour, and the emergence of a new spatial division of the workforce within Portugal itself. As a consequence, migration patterns in the 1990s have helped to reinforce the uneven spatial pattern that underpins Portugal's economic development structure, exaggerating the influence of the urbanised coastal margin over the more isolated rural interior.

At the end of 1997, Portugal's legally resident foreigners (LRF) numbered in excess of 175,000, and represented an estimated 1.9 per cent share of the country's working population. While foreigners remain a small minority, two factors point to their growing importance. First, the resident foreign population is growing rapidly, with an increase of 95% over the period 1987-97 (INE, 1988; 1998). Second, many foreign immigrant groups tend to be spatially concentrated at both the regional and local scales. Consequently when large migrant groups gather in close proximity then so their influence, and more importantly the perceptions of their influence, can grow enormously within localised job markets. This process, in turn, is linked to the growing internationalisation of the Portuguese economy (particularly in services and industry), as well as to the entrenched nature of immigration (both legal and clandestine) in this part of southern Europe (Doomernik, 1997; King and Black, 1997).

Immigrants settling in Portugal reflect a mixture of ethnically and culturally diverse peoples and include Luso-Africans, south Americans and north Europeans, as well as growing numbers of Asians and east Europeans. Many are engaged in a range of jobs from the manual worker in the construction and domestic service sectors, through to highly trained professionals in the financial and medical service industries. In addition to the official immigrant labour market there is also a sizeable, but largely undocumented, presence of illegal immigrants. These clandestine migrants have also influenced the socio-economic development of the country but relatively little is known about their contribution.

With this background in mind, this chapter assesses the role of Portugal within the European Union's (EU) international migratory patterns, outlines the quantitative movements of immigrants into Portugal, and their spatial impacts upon national, regional and local labour markets. In conjunction with a resurgent literature on this topic, differences in the immigrant's roles in the legal and illegal sectors of economic activity are examined. Likely future trends, especially with respect to the implementation of immigration policy, are also outlined. Finally, it is argued that the duality of the labour market will remain an important factor in supplementing and rejuvenating the Portuguese economy in the future. It is important to emphasise that the subject matter of this chapter is an extremely grey area within Portuguese society. Features that are by their very nature, illegal, do not lend themselves to easy scientific investigation. Some of the findings, therefore, remain speculative and are at best a broad interpretation of the migration process as it is now affecting this westernmost part of Iberia.

International Population Mobility

Iberia is currently one of the main 'blue-border' (or sea-route) gateways to the EU for the entrance of illegal immigrants coming from north Africa (Economist, 1999, p.39). As signatories to the Schengen Accord (Convey and Kupiszewski, 1995), both Spain and Portugal offer potentially lucrative opportunities for labour migrants. Those who succeed can enter and travel around the EU, relatively unhindered, once residence has been established. The majority of academic attention has, however, focused upon Spain, and in particular, its relationship with Morocco, meaning that Portugal has remained somewhat neglected. Indeed, Portugal has rarely featured on the European Commission's agenda where the issue of immigration is concerned (Geokas, 1997). This is not entirely surprising given that the country has one of the smallest foreign populations (estimated at 1.6%) in the whole of the EU; a rate directly comparable to Greece (where LRFs represent 1.5% of the total population), but significantly less than other EU member states such as Luxembourg (32.7%), Belgium (11.3%), Austria (9.0%) and Germany (8.6%) (Eurostat, 1997, pp.86-87).

In spite of this small foreign presence Portugal has recently assumed a more important position within the European Union's migratory pattern, reflecting (albeit on a small-scale), the growing significance of a phenomenon called 'substitute' or 'shuffle' migration. This should not be confused with the notion of 'step' migration. Portugal is only rarely used as a stepping stone whereby immigrants enter the country and utilise it as a starting point before moving on further into the EU.[1] Indeed, very little immigration into Portugal is via sea-routes, with the vast majority being facilitated by air travel. Step migration is more common in Spain and Italy, where north African immigrants, in particular, look to use their host countries as starting points for travel on towards countries such as France and Germany (Veiga, 1998).

Traditionally, the Portuguese population have emigrated (Amaral, 1993; Brettell, 1993; Cavaco, 1995; Baganha, 1998a). While Figure 5.1 shows that this trend has continued, the country now finds itself in an important strategic position with its own niche market with respect to global migration patterns. For example, Portugal in the 1990s imported foreigners from both the developed and less-developed parts of the world economic system. Figure 5.1 also shows that there are significant intercontinental, and intra-European Union movements of people. Consequently, this in-migration helped to revive and rejuvenate the domestic labour market, with many of the new immigrants replacing or substituting, and in some cases supplementing, the national emigrant labour force in several important sectors of economic activity.

Table 5.1 illustrates the varied nature of Portugal's migratory balances in relation to the origins and destinations of migrants by world regions (INE, 1994-98). While Portugal at the international scale exported around 90% of emigrant Portuguese to northern Europe, it imported workers from Africa, the EU, South America, and Asia. In a long-established trail, Figure 5.1 confirms that Portuguese emigrants continued to flow towards the traditional destinations of the United Kingdom, France, Germany, and especially Switzerland. Within Europe it is estimated that there are 1.3 million Portuguese emigrants residing outside of their homeland with most (up to 800,000) living in France (Malaurie and François, 1998). Immigrants, on the other hand, arrive in Portugal particularly from the Cape Verde Islands, Angola, Guinea-Bissau, and Brazil (Engerman and Neves, 1997).

This process of substitution as some migrants move into, and others move out of the country can be explained by Portugal's position as a semi-peripheral country within the world economic system. In the international division of both labour and salary levels, emigrant Portuguese continue to leave because they can 'shuffle up' the social pyramids of employment and wages that are found in the more advanced EU nations. In short, they may travel and take up similar jobs to the ones they have left behind, but they will earn (in relative terms) much more for the same type of work. Likewise, immigrants to Portugal, especially those from Lusophone Africa, continue to

Table 5.1 Portugal's International Migration Balance by World Region, 1992-96

World Region	Permanent[1] Emigration	Permanent[2] Immigration	Balance
European Union	35,666	10,462	- 25,204
Other European	15,318	308	- 15,010
Africa	4,047	27,485	+ 23,438
North America	1,682	1,353	- 329
Central & South America	–	5,310	+ 5,310
Asia	–	2,371	+ 2,371
Total	56,713	47,289	- 9,424

Sources: INE, Various, 1993-1998

Note: [1] Based upon individuals emigrating for a period of greater than one year
[2] Based upon individuals legally registering with the Foreigners and Frontiers Service (SEF)

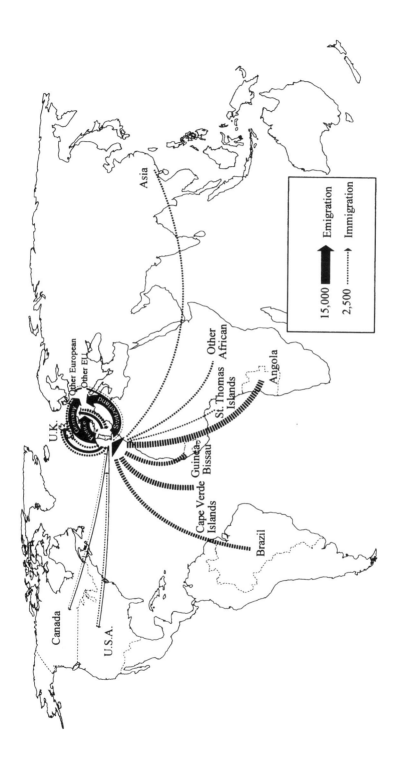

Figure 5.1 Portugal's Migratory Flows by Main Areas of Origin and Destination, 1992–96

enter and gain a vital first foothold near the bottom of this same remunerative pyramid (Fonseca, 1996; 1997). By utilising the higher earning potential and the greater remittance value of wages paid in Portugal, many can rationalise the relative hardship that they face. Their living conditions may be of a generally poor standard and their salaries low in EU and Portuguese terms. However, their willingness to accept such a vulnerable position on this pyramid means that their capacity for self-exploitation is relatively high. Moreover, it is an important conditioning factor in the relationship between immigration and the labour market in Portugal. Many immigrants will accept a semi-slave-like existence in Portugal in contrast to a life of poverty and/or the threat of physical violence in their homeland (Eaton, 1993).

Portugal's 'New Immigration'

Southern Europe, and in particular, the countries of Italy, Spain and Greece have undergone a remarkable change of status since the middle of the 1970s. Each has changed from being a region of mass emigration to one of mass immigration (King and Rybaczuk, 1993). There is a consensus that the same

Table 5.2 Portugal's International Migration Balance, 1992-97

Year	Permanent[1] Emigration	Permanent[2] Immigration	Balance
1992	22,324	8,370	- 13,954
1993	15,562	9,245	- 6,317
1994	7,845	25,480	+ 17,635
1995	8,109	11,243	+ 3,134
1996	9,598	4,596	- 5,002
1997	–	2,351	–
Total (1992-96)	63,438	58,934	- 4,504

Sources: INE, Various, 1993-1998

Note: [1] Based upon individuals emigrating for a period of greater than one year
[2] Based upon individuals legally registering with the Portuguese Foreigners and Frontiers Service (SEF)

pattern has emerged in Portugal (Esteves, 1991; Rocha-Trindade, 1995; Guibentif, 1996), although as Table 5.2 shows, it is wrong to assume that Portugal has become a net in-migrator of population. The overall picture is more complicated and the yearly fluctuations between 1992 and 1997, while hinting at a turnaround, confirmed little apart from the intractability of the migratory phenomenon. Table 5.2 suggests that a watershed was reached in 1994 with Portugal's migration balance sheet changing to one of net inward movement. The figure of +17,635 is, however, distorted because it includes large numbers of immigrants officially registered as a result of the government's first regularisation amnesty granted in 1992/93.

Overall, levels of emigration from Portugal are now much lower than in the peak years of the early 1970s,[2] but they remain a constant feature. There has been a considerable slowdown in the rate of out-migration, moving from 4.3 per thousand in 1986, to 1.0 per thousand of the Portuguese population in 1996 (Eurostat, 1998). Nevertheless, and in a continuation of the established pattern, young emigrant Portuguese are still being attracted to northern Europe. In the 1990s, however, the profile of the typical emigrant gradually changed from the relatively unskilled labourer to the semiskilled and professional worker. This is due partly to the higher standards of education and linguistic skills being attained by the Portuguese labour force, and partly to the greater demands for specialised and footloose workers being made by the north European labour markets (Machado, 1997).

Equally, immigration into Portugal has grown in importance since the overthrow of the country's authoritarian regime in April 1974, and in a direct response to burgeoning job opportunities and improved quality of life as Portugal has undergone a far reaching modernisation process (Williams, 1992; Syrett, 1997; Corkill, 1999). The significance of the foreign immigrant in Portugal is clearly a post-Revolution phenomenon. During the 1960s, Portugal housed a small foreign community of around 30,000, most of whom (approximately 67%) were European. After 1974, immigration into Portugal rose sharply and was due mainly to an influx of Luso-African workers, many of whom travelled in order to help with the physical expansion of the host economy. In short, many immigrants from the PALOPS (Portuguese speaking African countries) travelled to work in the construction trades and building industry, and indeed, thousands continue to do so.

Between 1977-97, Portugal's legally resident foreign community quadrupled in size (Barreto, 1996; INE, 1998) to reach an official total of more than 175,000 (see Table 5.3). Much of the growth can be explained by the close cultural and linguistic ties that exist between Portugal and the

Lusophone nations. Culturally it continues to be a relatively easy move to make for migrants, especially those travelling from Angola, Mozambique, the Cape Verde Islands, Guinea-Bissau and the Saint Thomas Islands, into Portugal. Movement has been further aided by a relatively lax approach on the part of the Portuguese state to limiting the numbers of foreigners entering the country. Relatively low unemployment levels have also meant less competition between domestic and foreign workers for jobs. Equally, Portugal's relatively peaceful social climate and lack of overt racismtowards immigrant communities continues to make it an attractive destination (Corkill, 1996).[3] As a result, the Portuguese government has been generally reluctant to intervene where immigrants are concerned. This stance stands in stark contrast to elsewhere in the EU, where other countries have gradually introduced stricter controls on both economic immigrants and asylum seekers (Geddes, 1995; Baldwin-Edwards, 1997).

Illegal Immigration

Because Portugal was seen as a country with few restrictions on official immigrants,[4] a second feature - illegal immigration - has also emerged as an important factor (Montagnè-Villette, 1994). King (1998, p.293), for example, suggested that the total number of immigrants in Portugal in the mid-1990s was between 200-250,000. The difficulty in defining a precise figure was due to the weak statistical hold that Portugal has over its foreign population,[5] and because of a relatively high proportion of immigrants who have entered and now live illegally in the country. On the basis of the figures given above, it is possible to derive a tentative figure of between 25,000-75,000 illegal immigrants living in Portugal. Speaking publicly in May 1999, Armando Vara from the Secretary of State's Office suggested a figure of 10,000 illegals, but one of the main non-governmental organisations for supporting immigrants, *Olho Vivo*, put the figure at 30,000. The discrepancies in the estimates reflect the lack of control both statistically and in terms of the highly sophisticated underground networks that now operate both within and outside of the country. These organisations, based mainly in African and eastern European countries, view Portugal as a lucrative destination for their nefarious activities and have been quick to establish themselves.

A common pattern is for illegals to arrive by aeroplane (perhaps in possession of tourist or student visas), and then be integrated (i.e. they 'disappear') into an informal labour market via the use of intermediaries and

Table 5.3 Changes in the Numbers of Legally Resident Foreigners in Portugal, 1987-97

Country of Origin	LRF Totals			Variation (No. & %)					
	1987	1992	1997	1987-92		1992-97		1987-97	
Cape Verde Islands	26,565	31,127	39,789	4,562	+17	8,662	+28	13,224	+50
Angola	4,187	6,601	16,296	2,414	+58	9,695	+147	7,922	+189
Guinea-Bissau	2,688	5,808	12,785	3,120	+116	6,977	+120	10,097	+376
Mozambique	2,600	3,574	4,426	974	+37	852	+24	1,826	+70
St. Thomas Islands	1,625	2,519	4,304	894	+55	1,785	+71	2,679	+165
Other African countries	1,173	2,408	4,117	1,235	+105	1,709	+71	2,944	+251
The United Kingdom	6,577	9,284	12,342	2,707	+41	3,058	+33	5,765	+88
Spain	6,985	7,734	9,806	749	+11	2,072	+27	2,821	+40
Germany	3,862	5,404	8,345	1,542	+40	2,941	+54	4,483	+116
France	2,673	3,674	5,416	1,001	+37	1,742	+47	2,743	+103
The Netherlands	1,408	2,010	3,149	602	+43	1,139	+57	1,741	+124
Other European countries	4,171	5,390	10,689	1,219	+29	5,299	+98	6,518	+156
Brazil	7,830	14,048	19,990	6,218	+79	5,942	+42	12,160	+155
Venezuela	4,738	4,910	3,783	172	+4	-1,127	-23	-955	-20
Other Central/South American	441	1,002	1,501	561	+127	499	+50	1,060	+240
United States of America	6,184	7,321	8,364	1,137	+18	1,043	+14	2,180	+35
Canada	2,266	2,109	2,209	-157	-7	100	+5	-57	-2
Asian countries	3,124	4,769	7,192	1,645	+53	2,423	+51	4,068	+130
Oceanic countries	313	391	487	78	+25	96	+25	174	+56
Portugal	89,778	121,513	175,263	31,735	+35	53,750	+44	85,485	+95

Sources: INE, 1988 p.201: INE, 1993 p. 182; INE, 1998 p. 155

organised employers' networks (Champion, 1998). These types of immigrant offer flexibility; they can be 'hired and fired' easily, incur few labour costs (i.e. welfare contributions are negligible on the part of employers), and can be controlled by 'agents' who may hold their (often false) passports and documentation as collateral against desertion. Many are unskilled labourers who have entered Portugal's burgeoning black economy (Miguelez-Lobo,1990; Baganha, 1998b). In turn, many have found employment in the civil construction, tourism, and domestic service industries (Campani, 1993), where the use of numerous layers of subcontractors makes their detection and the prosecution of their 'real' employers virtually impossible (Malheiros, 1999). In Lisbon, many illegals have settled in the *bairros de lata* (shanty-towns) in the capital's suburbs and along major transport arteries around the city (Eaton, 1993).[6] In spite of the precariousness of their positions economic activity rates among illegals are high; a function of an extremely high and sustained demand for labour in areas and sectors that are now in the vanguard of Portugal's modernisation process.

Immigrant Sources

Within the overall growth of the legally resident foreign community in Portugal, immigrants' origins remain diverse. Table 5.3 indicates origins reflect a multicultural range of Luso-African, north European, south, and north American, as well as Asian and Oceanic sources (Corkill and Eaton, 1998). Of the smaller communities, the Asians are among the most important in terms of recent growth rates. In the period, 1992-97, legally resident foreigners from Asia increased in numbers at twice the rate of the Cape Verdean and Mozambican groups. Part of the reason for this was the influx of Portuguese passport holders from the former enclave of Macau in the Far-East. There are also growing numbers travelling (often illegally and with the alleged involvement of organised crime gangs) from mainland China. Many can be found working throughout Portugal (but especially in Lisbon) in restaurants or in shops selling traditional Chinese products to their compatriots.

This recent upsurge in numbers from Asian source regions is significant because Portugal's immigrant profile is dominated by one group, the Cape Verdeans (França, 1992). In 1997 they numbered just under 40,000 (or 23% of the total), and were twice the size of the next largest group from Brazil. They have traditionally entered Portugal to fill the gaps in the lower end of

Box 5.1 The Regularisation of Portugal's Immigrant Community

In the first half of the 1990s, Portugal became more aware of its obligations to the Schengen Accord and its responsibilities for the welfare of its labour force under the European Social Charter. Part of its response was to offer amnesties to resident, but undocumented, clandestine immigrants. Under the first extraordinary legalisation programme in 1992/93, some 39,166 illegals made formal requests for their status to be legalised (Kobayashi, 1998). However, it quickly became clear that many who worked in the informal economy were reluctant to compromise their earning potential given that a residence permit brought with it a requirement to pay income tax. Equally, some employers were reluctant to lose the competitive advantage they derived from employing cheap labour with few social security obligations. As thousands failed to respond it became difficult for the Portuguese state to monitor its foreign population and to enforce border controls.

Consequently in 1996 a second amnesty was granted and resulted in the identification of approximately 35,082 irregular migrants who were approved. A further 5,000 settlers were identified but for various reasons (i.e. possession of a criminal record, no job, and other irregularities) were not accepted. Figures from the *Serviço de Estrangeiros e Fronteiras* (SEF – Foreigners and Frontiers Service) confirmed that most of these approvals were from Angola (26% of the total numbers legalised), Cape Verde (19%) and Guinea-Bissau (15%) (Baganha, 1998b, p.371). Such relatively large numbers also confirm an earlier suggestion that most illegal immigrants entering Portugal tend to settle for long periods of time and show few propensities for either moving on to other countries, or indeed returning to their countries of origin.

It is unclear, however, precisely how many undocumented immigrants now remain in Portugal. Malheiros (1998, p.9) claims that they remain an important feature and there are several factors, over and above those already mentioned, which account for this uncertainty. First, there is the indeterminate number of recent arrivals (i.e. those settling illegally since the last regularisation took place). Second, there are those who were unable to formally present their documents in 1996. Third, there are those who were legally forbidden to work (i.e. asylum seekers) but who may have drifted into the informal sector in order to find jobs and support themselves as they await a decision upon their asylum request. Finally, there are those who simply have no desire to place themselves on the government's books, preferring instead to reap what they perceive to be the advantage of illegal status.

the job market left by young emigrant Portuguese males (Filho, 1996). Indeed, this group has a long association with Portugal, dating back to the 1960s when economic expansion, intensive out-migration, and the military manpower requirements of the colonial wars led to acute labour shortages in Portugal. The Cape Verdeans were the first to travel to Portugal in significant numbers in order to take advantage of the job opportunities that were opening up; a pattern that continued across the 1990s. However more recently, Cape Verdean immigrants have tended to be encouraged to leave as much by poverty and civil instability at home, as by economic opportunity in Portugal (Brookshaw, 1992).

In relative terms the dominance of the Cape Verdean community is in decline. As has been mentioned, other Asian and Lusophone groups have accelerated their growth rates. For example, between 1987-97, immigrant groups moving in from Guinea-Bissau and Angola have grown significantly. In the former case, the Guinea-Bissaun community grew by a remarkable 376%. However, much of the increase was less the result of Portugal's popularity for new immigrants from Guinea-Bissau and Angola, and more the results of regularisation amnesties granted in 1992/3 and again in 1996 (Kobayashi, 1998) (see Box 5.1).

Table 5.3 confirms that the Guinea-Bissaun immigrant community is now the fourth largest group resident in Portugal (7% of the total), preceded by, in second place, the Brazilians (11%), and in third, the Angolans (9%). What is unusual about the Guinea-Bissaun immigrants is that, as Machado (1997, pp.24-25) points out, they are often reasonably well-educated, entering the teaching as well as scientific and technical professions in Portugal. They do not, therefore, readily conform to the common image of unskilled, Luso-African immigrant workers. The picture consequently becomes more complicated as the level of analysis changes and the focus shifts to the occupational structures of different national groupings. Nevertheless, the over riding significance of the Lusophone countries, and the facility that is provided by a common language and culture in helping establish strong migration channels, remains paramount in explaining this particular pattern.

Labour Market Change

The settlement of Portugal's immigrant communities exhibits a strong spatial pattern characterised by a clear regional dimension and a focus upon Lisbon. This pattern (see Figure 5.2) helps to further amplify the underlying

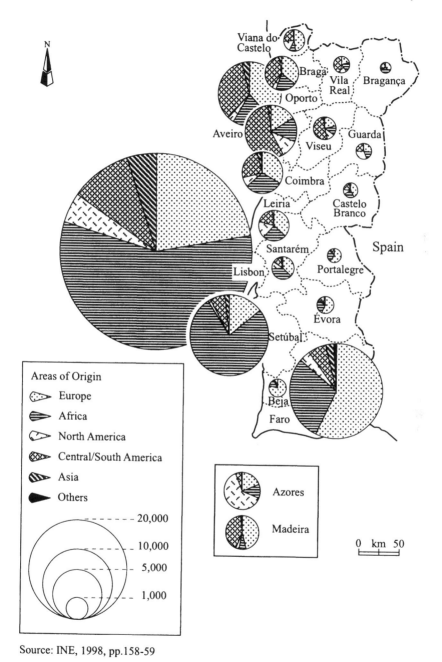

N

Areas of Origin
- Europe
- Africa
- North America
- Central/South America
- Asia
- Others

Azores

Madeira

0   km   50

Spain

Viana do Castelo
Braga
Vila Real
Bragança
Oporto
Aveiro
Viseu
Guarda
Coimbra
Leiria
Castelo Branco
Santarém
Portalegre
Lisbon
Setúbal
Évora
Beja
Faro

Source: INE, 1998, pp.158-59

Figure 5.2 Spatial Distribution of Legally Resident Foreigners in Portugal by Main Area of Origin, 1997

unevenness of the economic development process within Portugal. Most immigrants have converged upon the more modern, urbanised and industrialised western coastal margin and live in the band of *distritos* (counties) running from Braga in the north west, southwards through Oporto, Aveiro, Coimbra, Leiria, Lisbon, and on to Setúbal (an area that is sometimes termed the BOALLS axis). Immigrants tend to live in and around each of these major cities and have also been attracted into Faro in the far south and, to a lesser extent, towards the two autonomous island regions of the Azores and Madeira. These are the zones where the majority of jobs are, and when coupled with the received wisdom gleaned from established foreign communities, the BOALLS axis has quickly become a self-perpetuating focus of attraction for immigrants. The peripheral interior zones to the north and east which straddle the Spanish border are largely ignored by immigrants. Consequently, the west-south/north-east division with respect to the malapportionment of inflows of immigrants helps to reinforce the spatial unevenness that underpins Portugal's demographic and economic development at the present time.

The strong pull of the littoral margin and particularly that of the capital can be seen in the fact that Lisbon is home to the majority (55%) of all Portugal's immigrants, although foreigners still represent less than 5% of the capital's total population. The AML is the zone with the highest concentrations and variety of immigrants. In 1997, the Lisbon Metropolitan Area (AML) housed approximately 28% of the country's total population, but some 67% of all African, and 69% of all Asian immigrants to be found in the country. In addition, 46% of all north Americans and 43% of Europeans are now settled in the capital. This spatial clustering is related, in part, to the strong social networks that have developed, particularly among low social status, Luso-African workers, many of whom have settled in the shanty settlements found in the capital's suburbs (Branco, 1999). For Asian immigrants the channelling effect is also related to the recent upsurge in commercial ventures (i.e. Indian furniture shops/electrical products, and Chinese/Macanese restaurants) in previously unexploited niche markets. Again these are focused upon Lisbon and the suburbs of Mouraria and Martim Moniz, in particular (Malheiros, 1996a; 1996b; 1997). At the local scale, immigrants congregate together in demonstrations of ethnic solidarity and the public perception among Portuguese of their relative influence at this level, can sometimes outstrip their significance in absolute terms.

Table 5.4 shows that the overt presence or visibility of foreigners is greatest in the *distritos* of Faro (6.5%), followed by Lisbon (4.7%) Setúbal

(2.2%), and the Azores (1.1%).[7] In most other cases the foreign presence is negligible, and surprisingly even in the industrialised centre of Oporto immigrant 'visibility' is low at just 0.6%. In relative terms, therefore, Faro stands out as the major destination for foreigners settling in Portugal. Indeed, while the Algarve is home to less than four out of every one hundred of the Portuguese population, it provides residence for thirteen out of every 100 legally resident foreigners in the country.

The district of Faro has been 'internationalised' by relatively large numbers of British, German and Dutch residents (see Table 5.5). Collectively, these three groups accounted for 44% of all foreigners in this region in 1997. This north European influence is related to the importance of the Algarve as a venue for consumption upon retirement for expatriates (Williams, et al, 1997; Williams and Patterson, 1998; see Chapter 4). Equally, there is a large group of immigrant workers servicing the Europeanisation of tourism and the expatriate residents in this southern Portuguese region (Eaton, 1998, p.51). There is also a significant level of Cape Verdean settlement, reflecting the ubiquitous nature of their presence, but more importantly, their involvement in the region's construction of tourist accommodation. Lisbon and Setúbal, on the other hand, both have large Luso-African (Cape Verdean and Angolan), and South American (Brazilian) populations. Here, the situation is more complicated with workers at the lower end of the socio-economic pyramid (in sectors such as construction, public works and domestic service), alongside professional workers who lie at the opposite end of the spectrum (in sectors such as medicine, the media, and sport).

Given that the main motivation for moving to Portugal is to find a job (see Table 5.6), most immigrants can be classified as economic or labour migrants. The spatial focus of immigrants in the Metropolitan Area of Lisbon therefore reflects how Portugal's economic growth in the 1990s has centred on this area, particularly through major civil construction and public works programmes. Examples include, the building of the World Expo '98, the Vasco da Gama Bridge, several large city centre underground car parks, the new rail link between Lisbon and the south bank of the River Tejo, and the extensions to the metro and suburban rail networks. Private developments such as the new Colombo shopping centre (the largest of its type in Iberia) have also placed heavy demands upon the supply of labour in the capital. Indeed, this pressure has laid many workers open to exploitation (caught up in a long hours, low pay syndrome), and vulnerability via dangerous working conditions. This situation has further reinforced the

Table 5.4 Proportions of Legally Resident Foreigners by Host Area, 1997

Host County/ Autonomous Region	Total LRF Population	% of Total LRF	Total Resident Population[1]	% of Total of Resident Population	% LRF/ Head of Resident Population
Aveiro	6,899	3.9	660,807	6.6	1.0
Beja	777	0.4	171,132	1.7	0.4
Braga	2,854	1.6	755,673	7.6	0.4
Bragança	263	0.1	159,387	1.6	0.2
Castelo Branco	538	0.3	217,001	2.2	0.2
Coimbra	4,294	2.4	432,117	4.3	1.0
Évora	794	0.4	175,390	1.8	0.4
Faro	22,407	12.8	344,818	3.5	6.5
Guarda	652	0.4	190,046	1.9	0.3
Leiria	2,325	1.3	430,413	4.3	0.5
Lisbon	96,759	55.2	2,068,661	20.8	4.7

Table 5.4 (cont.)

Host County/ Autonomous Region	Total LRF Population	% of Total LRF	Total Resident Population[1]	% of Total of Resident Population	% LRF/ Head of Resident Population
Portalegre	507	0.3	135,510	1.4	0.4
Porto	10,669	6.1	1,657,916	16.6	0.6
Santarém	1,229	0.7	449,328	4.5	0.3
Setúbal	16,203	9.2	719,719	7.2	2.2
Viana do Castelo	1,143	0.6	252,559	2.5	0.4
Vila Real	642	0.4	238,656	2.4	0.3
Viseu	1,204	0.7	405,889	4.1	0.3
Açores	2,765	1.6	240,172	2.4	1.1
Madeira	2,339	1.3	255,960	2.6	0.9
PORTUGAL	175,263	100.0	9,961,165	100.0	1.8

Source: INE 1998, pp. 158-159. Note: [1] Author's estimate.

Table 5.5　Spatial Distribution of Legally Resident Foreigners by Main Areas of Origin, 1997

Main Areas of Origin (% of total)

Host County/ Autonomous Region	Total	%	Africa	%	Europe	%
Aveiro	6,899	3.9	1,049	15.2	1,079	15.6
Beja	777	0.4	98	12.6	608	78.2
Braga	2,854	1.6	586	20.5	992	34.8
Bragança	263	0.1	94	35.7	90	34.2
Castelo Branco	538	0.3	128	23.8	225	41.8
Coimbra	4,294	2.4	1,304	30.4	1,430	33.3
Évora	794	0.4	211	26.6	453	57.0
Faro	22,407	12.8	6,453	28.8	12,781	57.0
Guarda	652	0.4	113	17.3	173	26.5
Leiria	2,325	1.3	594	25.5	842	36.2
Lisbon	96,759	55.2	55,088	56.9	21,429	22.1
Portalegre	507	0.3	142	28.0	305	60.1
Porto	10,669	6.1	2,178	20.4	4,082	38.3
Santarém	1,229	0.7	457	37.2	451	36.7
Setúbal	16,203	9.2	12,324	76.0	2,279	14.0
Viana do Castelo	1,143	0.6	118	10.3	543	47.5
Vila Real	642	0.4	131	20.4	130	20.2
Viseu	1,204	0.7	160	13.3	258	21.4
Açores	2,765	1.6	331	11.2	524	18.9
Madeira	2,339	1.3	158	6.7	1,073	45.9
Portugal	175,263	100.0	81,717	46.6	49,747	28.4

attractiveness of illegal workers. Illegals who have no redress to trade union protection, no basic minimum wage nor any insurance against injury, disablement or death, cost their employers very little. Wages can be as low as the owner sees fit and indirect costs, such as social security payments, are bypassed or conveniently ignored.

Among immigrants, occupational activity rates can be as high as 50%, compared to only 45% for the Portuguese population as a whole (INE, 1998;

(Table 5.5/cont.)

Main Areas of Origin (% of total)

Host County/ Region	Central/ South America	%	North America	%	Asia	%	Other	%
Aveiro	3,855	55.9	760	11.0	141	2.0	15	0.2
Beja	49	6.3	17	2.2	5	0.6	0	–
Braga	1,058	37.1	125	4.4	86	3.0	7	0.2
Bragança	53	20.1	9	3.4	14	5.3	3	1.1
Castelo Branco	110	20.4	52	9.7	20	3.7	3	0.6
Coimbra	1,070	24.9	311	7.2	150	3.5	29	0.7
Évora	85	10.7	23	2.9	17	2.1	5	0.6
Faro	1,390	6.2	832	3.7	745	3.3	206	0.9
Guarda	153	23.4	197	30.2	15	2.3	1	0.1
Leiria	442	19.0	398	17.1	37	1.6	12	0.5
Lisbon	9,976	10.3	4,913	5.1	4,992	5.1	361	0.4
Portalegre	39	7.7	14	2.8	6	1.2	1	0.2
Porto	3,595	33.7	328	3.1	463	4.3	23	0.2
Santarém	145	11.8	129	10.5	40	3.2	7	0.6
Setúbal	941	5.8	232	1.4	371	2.3	56	0.3
Viana do Castelo	270	23.6	185	16.2	25	2.2	2	0.2
Vila Real	296	46.1	82	12.8	3	0.5	0	–
Viseu	618	51.3	139	11.5	27	2.2	2	0.2
Açores	139	5.0	1,744	63.1	19	0.7	8	0.3
Madeira	990	42.3	83	3.5	16	0.7	19	0.8
Portugal	25,274	14.4	10,573	6.0	7,192	4.1	760	0.4

Source: INE 1998, pp.158-59

Eurostat, 1997) (see Table 5.6). In terms of the occupational structure of the economically active immigrant population, the largest groups are composed, firstly, of manual workers (48%), and secondly, of professional employees (24%). Although these figures do not include illegal workers, it is evident that almost three quarters of foreigners work in these two segments of the Portuguese labour market. The largest group, the unskilled manual labourers (coming mainly from Africa and the Cape Verde Islands) fill the most difficult and low paid jobs in civil construction, hygiene, transport, domestic services and manufacturing industry. At the opposite end of this labour market hierarchy are the relatively well paid managers and self-employed professionals from Europe, South America, and especially Brazil (see Box 5.2). Despite the differences in their relative position within the Portuguese labour market these immigrant groups are linked together via the European wide segmentation of labour markets. With de-industrialisation in the north of Europe and a pattern of relocation of industry (especially of branch-plants) towards the south of Europe together with increased foreign direct investment, so the demands for specific types of workers have grown.

Table 5.6 Occupational Structure of Legally Resident Foreigners in Portugal, 1997

Nationality	Active Pop^n.	Professions						
		0-1	2	3	4	5	6	7-8-9
European	26,273	11,053	3,688	1,327	3,300	1,535	477	4,893
(UK)	5,482	2,615	1,069	327	429	386	95	921
African	43,299	2,298	350	1,272	1,439	4,280	294	33,366
(Cape Verde)	22,073	361	15	650	169	1,956	130	18,792
North America	3,694	2,197	230	84	120	7	185	871
(USA)	3,188	2,049	193	66	74	1	145	660
South American	11,000	4,957	493	711	1,361	734	89	2,655
(Brazil)	9,671	4,459	425	684	1,171	695	69	2,168
Asian	3,310	691	300	83	948	789	31	468
Others	317	115	23	10	37	13	12	109
Total	87,893	21,309	5,084	3,487	7,205	7,358	1,088	43,362

Source: INE 1998, p.160

Key to Professions

0-1	Professional, scientific, technical, self-employed, and related occupations
2	Executive, administrative and managerial occupations (public and private sectors)
3	Clerical and office workers
4	Commercial sales persons
5	Service sector workers
6	Agriculture, animal husbandry, forestry, fisheries, hunting and related workers
7-8-9	Industrial production and related workers (in manufacturing), transport equipment, and manual labourers (in building and construction trades)

Hence, the importation into Portugal of manual operatives and construction workers has proceeded side by side with the recruitment of professional managers, scientists, and technical staff (Ferrão and Vale, 1995).

Box 5.2 Immigration From Brazil

The Brazilian immigrant communities entering Portugal in the middle to late 1990s present an interesting case. As foreign investment and business opportunities grew in Portugal, so economic recession and social instability in Brazil encouraged emigration. Given the strong historical link between the two countries, a two-way migration channel opened up. This linkage was facilitated by a shared culture, language and information trail, and supported by family networks on both sides of the Atlantic Ocean. In a small echo of the traditional flow of Portuguese to Brazil whereby an estimated 1.2 million have settled in south America, a significant number of Brazilians 'returned' to Portugal. Between 1987-97, for example, more than 12,000 legally resident Brazilians settled in Portugal, bringing the official community up to a level of almost 20,000.

What is particularly notable is that the majority (58%) of the economically active Brazilian population are either professional workers or sales people, making their mark in sectors as diverse as medicine, road construction, soccer, the media and TV. Equally important is the fact that the Brazilian community grew at a rate of 155% between 1987-97, faster than for any of the European Union immigrant groupings. In spatial terms, the Brazilians are also unusual in that they are prominent in the north west of the country, particularly around Vila Real, Braga, Oporto, Aveiro, and Viseu. This spatial concentration has emerged because many of the newer immigrants represent the counter-wave to earlier Portuguese out-migration, which originally flowed from around the Oporto region (Fonseca, 1997). Classical two-way information channels (Rowland, 1992) have been established and Brazilians have now returned in some strength to the original source areas of their parents and relations.

Although there is evidence of a degree of animosity directed towards Brazilian groups by some Portuguese who accuse them of involvement in criminal activities such as prostitution, money laundering and drugs supply (Lamb, 1995; Rocha, 1999), overwhelmingly Brazilians have been assimilated into Portuguese society with relatively few problems. Indeed their impact, especially upon popular culture - Portuguese television is dominated by Brazilian-made soap operas - continues to grow.

Conclusions

The pattern of emigration from, and immigration into Portugal, continues to shape the country's demographic structure, and more importantly, the socio-economic development process. While foreign residents remain a small group, representing less than two per cent of the national population, their presence at the regional and local scales often transcends their minority status. Immigrant labour is an integral, if small, part of Portugal's contemporary labour market. Immigrants, both legal and clandestine, are substituting, and in some cases supplementing, the domestic workforce, particularly in situations where demand for labour continues to rise.

Immigrants are a significant component of what has become Portugal's dual labour market. As this chapter has shown, they are significant players in the formal, and dominant in the informal, sectors of economic activity. They occupy a range of employment positions, from the professional and skilled worker to the manual and semi or low-skilled labourer. The origins of immigrants are diverse, but there tends to be a relationship between nationality and their position within Portugal's socio-economic pyramid. In its simplest manifestation, Cape Verdeans and Angolans occupy the lowest points with many working in (relatively) low paid, menial jobs. Near the top of this pyramid sit the north European professionals, who are well-paid, and the retirees who are enjoying a relatively good quality of live. Just below them come two Lusophone groups, the Brazilians and increasingly the Guinea-Bissauns, who have moved away from traditional low status jobs to enter more prestigious and lucrative jobs in the professional, scientific, teaching and managerial classes.

Regularisation has helped to stabilise the position of legally resident foreigners in Portugal but as long as cheapness and flexibility continue to be offered by illegal immigrants, then they will continue to be employed. In order to eradicate their presence much depends upon the attitude of the Portuguese state. To date, government appears to have been ambivalent; privately acknowledging the immigrants' important contributions to the modernisation process whilst reluctantly recognising the need for some stricter control of illegal flows. However, in the future the Portuguese government may be required to be more active. The European Commission is likely to push further for stronger control of its external borders and internally attitudes towards Portugal's immigrants may harden.

The SEF has recently demonstrated a willingness to increase their inspections of illegal immigrant communities. Significantly, these operations (mainly carried out in the Algarve in 1999) netted labourers from a major new source area - eastern Europe - notably from Moldova, the Ukraine and Rumania. In a clear echo of the use of intermediaries among illegal entrants from Lusophone African states, many of the eastern European clandestines are now being assisted by illegal, but internationally based agencies. Most of these criminal organisations are providing false documentation, charging high prices,[8] and pocketing the commission (Milhazes, 2000). Clearly, these sophisticated networks are increasing their influence and since prosecution and deterrents are rare, then they are likely to continue to grow. If, on the other hand, the clampdown from the SEF continues, then it will make all immigrant groupings' efforts to settle in Portugal a much harder proposition, whilst stronger attempts to regulate the informal economy will lead to an increased rate of arrest of both domestic and foreign workers. Nevertheless, it is unlikely that expulsions will reflect more than the tip of the iceberg of those illegals trying to enter Portugal (irrespective of those illegals already living and working in the country). As a consequence, while the demand for cheap, flexible workers remains among employers, then many immigrants will be willing to provide them with their labouring skills.

There appears little to stop the foreign community in Portugal from continuing to expand in the near future. Indeed, if the projected influx of Chinese/Macanese materialises and the recent wave of immigrants from eastern Europe is left unchecked, then current growth rates (of approximately +9.5% p.a.) will accelerate rapidly. Lisbon is set to continue to be the focal point of this growth given its place at the hub of both the formal and informal economies. Consequently it is likely to reap the economic and social benefits of the diversity brought by an increasingly multi-cultural capital, but also the difficulties which arise from growing concentrations of ethnic minority populations. More broadly this spatial focus is likely to exaggerate still further the underlying pattern of uneven regional development within Portugal. As immigrants settle they indirectly reinforce the dominance of the western/southern coastal margin over the northern/eastern interior. Although the government appears to neither condone nor condemn their presence, it is clear that the contributions of Portugal's foreign immigrants will continue to increase and it would be unwise for their role to be ignored.

Acknowledgements

I am grateful for advice and assistance provided by David Corkill, Carlos Pereira da Silva and Allan Williams in the compilation of this chapter. Equally, I am indebted to information supplied by Virgínia Branco. However, all of the material, viewpoints and the errors they contain remain the responsibility of the author.

Notes

1   There are unconfirmed suggestions that some immigrant groups entering Portugal are more likely than others to move around the Schengen area once a valid residence permit has been attained. Settlers from India, Pakistan and Bangladesh are rumoured to be among this group of 'step' migrants, but they remain a small minority (Rocha, 1999).

2   1970 was the peak year for official Portuguese emigration with approximately 170,000 persons leaving in a twelve month period. The real figure taking into account illegal emigration was probably much greater. Current levels of official emigration are therefore a fraction (approximately 1/15th) of the highest rate.

3   While Portugal is not noted for experiencing the same levels of racial violence as the rest of the EU, there is evidence to suggest that racial harassment and/or discrimination does exist (Fekete, 1993). For example, the Police and the SEF (*Serviço de Estrangeiros e Fronteiras*) have both been criticised over their handling of immigrants at frontier posts (most notably at Lisbon's International Airport). In general however, levels of arrest and deportation of immigrants in Portugal are low, for example there were just 133 deportations in 1993 (Baldwin-Edwards, 1997, p.509). Immigrants are allowed to live essentially undisturbed by the authorities, although clampdowns have become more common. Rocha (1999), for example, claimed that in 1998 around 2,000 individuals were expelled, sent back, or detained pending deportation. Given that asylum seekers and political refugees only rarely target Portugal, then it is safe to assume that most of the 2,000 immigrants mentioned here were likely to be illegal labour migrants.

4   A legally resident foreigner (LRF) holds a residence permit. He or she can qualify on the basis of three conditions: the applicant should have been resident in Portugal for six years; should have a job; and must be judged 'socially adaptable/acceptable' by the SEF. The numbers recorded as LRFs in the tables accompanying the text include first destination immigrants, their spouses and dependants (INE, 1996, p.23). Illegal immigrants who are detained by the SEF and/or the police, are given two options. These include either a return to their native country at their own expense, or repatriation at the Portuguese state's expense, with the proviso that they are prohibited from exercising any activity in any Schengen Accord member state for a period of between three and five years.

5   Migration data in Portugal is fraught with errors and inconsistencies and not least with a lack of comparability. The problems are compounded by there being two different organisations charged with collating information. The National Statistics Institute (INE) collects emigration data, and the SEF collects statistics related to immigration (Machado, 1997, p.26-27).

6   In preparation for Expo-98, the AML's authorities embarked on an ambitious campaign to demolish and rehouse shanty dwellers in low-rent, purpose built accommodation. While shanty settlements remain, there has been some visible progress in terms of eradicating the immediate problem of the estimated 30,000 to 35,000 *barracas* (shacks) in the capital. However, wider questions related to whether or not the more acute problems of the shanties (drugs, crime, prostitution etc.) are simply being relocated rather than removed have still to be addressed.

7   In 1997, the Azores Islands contained more than 2,750 foreign residents, the vast majority of whom (63%) were registered as citizens from the USA. In a relatively small island community such large numbers can have a significant impact at the local scale. The Azores are popular with North American settlers as this flow of immigration is directly related to the presence of the American air force base at Lajes on the island of Terceira, as well as historical migration links between these islands and North America.

8   In 1999, one of the Chief Inspectors of the SEF suggested publicly that for an illegal emigrant from the former USSR, transit to Portugal with the help of an organised criminal network would cost approximately 200 *contos*. Such a figure ensured the provision of false documentation but did not include the cost of transport to Portugal (Rocha, 1999).

References

Amaral, J.F. do (1993), 'Portugal and the Free Movement of Labour', in J. da S. Lopes (ed) *Portugal and EC Membership Evaluated*, Pinter, London, pp.240-246.

Baganha, M.I. (1998a), 'Portuguese Emigration After World War II', in A. Pinto (ed) *Modern Portugal*, SPOSS, Palo Alto, Ca., pp.189-205.

Baganha, M.I. (1998b), 'Immigrant Involvement in the Informal Economy: the Portuguese Case', *Journal of Ethnic and Migration Studies*, vol.24, no.2, pp.367-385.

Baldwin-Edwards, M. (1997), 'The Emerging European Immigration Regime: Some Reflections on Implications for Southern Europe', *Journal of Common Market Studies*, vol.35, no.4, pp. 497-519.

Barreto, A. (ed) (1996), *A Situação Social em Portugal*, 1960-1995, ICS, Lisbon.

Branco, V. (1999), 'L'Immigration au Portugal: Une Nouvelle Immigration dans un Ancien Pays D'Emigration', unpublished Master's thesis, IEPPCSS, Paris.

Brettell, C.B. (1993), 'The Emigrant, the Nation, and the State in Nineteenth and Twentieth Century Portugal: An Anthropological Approach', *Portuguese Studies Review*, vol.2, part 2, pp.51-65.

Brookshaw, D. (1992), 'Islands Apart: Tradition and Transition', *Index on Censorship*, vol.6, pp.13-14.

Campani, G. (1993), 'Immigration and Racism in Southern Europe: the Italian Case', *Ethnic and Racial Studies*, vol.16, no.3, pp.507-35.

Cavaco, C. (1995), 'A Place in the Sun: Return Migration and Rural Change in Portugal', in R. King (ed) *Mass Migration in Europe*, Wiley, Chichester, pp.174-191.

Champion, T. (1998), 'Demography', in T. Unwin (ed) *A European Geography*, Longman, Harlow, pp.241-259.

Convey, A. and Kupiszewski, M. (1995), 'Keeping Up with Schengen: Migration and Policy in the European Union', *International Migration Review*, vol.29, no.2, pp.939-963.

Corkill, D. (1996), 'Multiple National Identities, Immigration and Racism in Spain and Portugal', in B. Jenkins and S.A. Sofos (eds) *Nation and Identity in Contemporary Europe*, Routledge, London, pp.155-171.

Corkill, D. (1999), *The Development of the Portuguese Economy: A Case of Europeanization*, Routledge, London.

Corkill, D. and Eaton, M.D. (1998), 'Multicultural Insertions in a Small Economy: Portugal's Immigrant Communities', *South European Society and Politics*, vol.3, no.3, pp.149-168.

Doomernik, J. (1997), 'Current Migration to Europe', *Tijdschrift voor Economische en Sociale Geografie*, vol.88, no.3, pp.284-90.

Eaton, M.D. (1993), 'Foreign Residents and Illegal Immigrants: Os Negros em Portugal', *Ethnic and Racial Studies*, vol.16, no.3, pp.536-62.

Eaton, M.D. (1994), 'Regional Development Funding in Portugal', *Journal of the ACIS*, vol.7, no.2, pp.36-46.

Eaton, M.D. (1998), 'Foreign Residents and Illegal Immigrants in Portugal', *International Journal of Intercultural Relations*, vol.22, no.1, pp.49-66.

Economist (1999), 'Europe's Smuggled Masses', *The Economist Newspaper*, 20 February, pp.39-40.

Engerman, S.L. and Neves, J.C. das (1997), 'The Bricks of an Empire 1415-1999: 585 Years of Portuguese Emigration', *Journal of European Economic History*, vol.26, no.3, pp.471-510.

Esteves, M. do C. (ed) (1991), *Portugal: Pais de Imigração*, IEPD, Lisbon.

Eurostat (1997), Eurostat Yearbook 1997: A Statistical Eye on Europe, 1986-1996, *Eurostat*, Luxembourg, June.

Eurostat (1998), Statistics in Focus: Population and Social Conditions, no.9, *Eurostat*, Luxembourg.

Fekete, L. (1993), 'Inside Racist Europe', in T. Bunyan *Statewatching the New Europe: a Handbook on the European State*, Statewatch, London, pp.154-172.

Ferrão, J. and Vale, M. (1995), 'Multi-Purpose Vehicles, A New Opportunity for the Periphery? Lessons from the Ford/VW Project (Portugal)', in R. Hudson and E.W. Schamp (eds) *Towards a New Map of Automobile Manufacturing in Europe?*, ESF, Strasbourg, pp.195-217.

Filho, J.L. (1996), 'Inmigrantes Caboverdianos en Portugal', Arbor, vol.154, no.607, pp.151-170.

Fonseca, M.L. (1996), 'Portugal in the International Migration System at the end of the 20th Century: Challenge and Change', paper presented at *Regional Conference of the International Geographical Union*, Prague, Czechoslovakia.

Fonseca, M.L. (1997), 'The Geography of Recent Immigration to Portugal', paper presented at *Conference on Non-Military Aspects of Security in Southern Europe: Migration, Employment and the Labour Market*, Santorini, Greece.

Fonseca, M.L. and Cavaco, C. (1997), 'Portugal in the 1980s and 1990s: Economic Restructuring and Population Mobility', in H.H. Blotevogel and A.J. Fielding (eds) *People, Jobs and Mobility in the New Europe*, Wiley, Chichester.

França, L. et al. (eds) (1992), A Comunidade Cabo Verdiana em Portugal, IED, Lisbon.

Geddes, A. (1995), 'Immigrant and Ethnic Minorities and the EU's Democratic Deficit', *Journal of Common Market Studies*, vol.33, no.2, pp.197-217.

Geokas, M.C. (1997), 'The EU and the Spectre of Uncontrolled In-Migration', *Journal of Political and Military Sociology*, vol.25, pp.353-362.

Guibentif, P. (1996), 'Le Portugal Face á L'Immigration', *Revue Européenne des Migrations Internationales*, vol.12, no.1, pp.121-138.

Instituto Nacional de Estatística (1988-1998), *Estatísticas Demográficas*, INE, Lisbon.

King, R. (1998), 'Post-Oil Crisis, Post Communism: New Geographies of International Migration', in D. Pinder (ed) *The New Europe: Economy, Society and Environment*, Wiley, Chichester, pp.281-304.

King, R. and Black, R. (eds) (1997), *Southern Europe and the New Immigrations*, Sussex Academic Press, Brighton.

King, R. and Rybaczuk, K. (1993), 'Southern Europe and the International Division of Labour: From Emigration to Immigration', in R. King (ed) *The New Geography of European Migrations*, Belhaven, London, pp.175-206.

Kobayashi, M. (1998), Narrative Construction of Reality: Portugueseness and Semiperipheral Space, *unpublished MA thesis*, University of Warwick.

Lamb, C. (1995), 'Brazilians Colonise Portugal', *The Sunday Times*, 29 October.

Machado, F.L. (1994), 'Luso-Africanos em Portugal: Nas Margins da Etnicidade', *Sociologia: Problemas e Práticas*, no.16, pp.111-34.

Machado, F.L. (1997), 'Contornos e Especifidades da Imigração em Portugal', *Sociologia: Problemas e Práticas*, no.24, pp.9-44.

Malaurie, G. and François, P.O. (1998), 'La Fierté Retrouvée des Portugais de France', *L'Européen*, no.8, pp.16-26.

Malheiros, J.M. (1996a), *Imigrantes na Região de Lisboa: Os Anos da Mudança*, Ed. Colibri, Lisbon.

Malheiros, J.M. (1996b), 'Communatés Indiennes á Lisbonne', *Revue Européenne des Migrations Internationales*, vol.12, no.1, pp.141-158.

Malheiros, J.M. (1997), 'Indians in Lisbon: Ethnic Entrepreneurship and the Migration Process', in R. King and R. Black. (eds) *Southern Europe and the New Immigrations*, Sussex Academic Press, Brighton, pp.93-112.

Malheiros, J.M. (1998), 'Foreign Workers in the Portuguese Labour Market: Facing Illegality and Vulnerability', *mimeo* (unpublished).

Malheiros, J.M. (1999), 'Immigration, Clandestine Work and Labour Market Strategies: the Construction Sector in the Metropolitan Region of Lisbon', *Southern European Society and Politics*, vol.3, no.3, pp.169-185.

Miguelez-Lobo, F. (1990), 'Irregular Work in Portugal', in Commission of the European Communities, *Underground Economy and Irregular Forms of Employment (Travail au Noir)*, Luxembourg, CEC.

Milhazes, J. (2000), 'Onda de Imigração Russa', *O Público*, 14 February.

Montagné-Villette, S. (1994), 'Mobility and Illegal Labour in the EC', in M. Blacksell and A.M. Williams (eds) *The European Challenge: Geography and Development in the European Community*, OUP, Oxford, pp.242-254.

Peixoto, J. (1996), 'Recent Trends in Regional Migration and Urban Dynamics in Portugal', in P. Rees et al. (eds) *Population Migration in the European Union*, Wiley, Chichester, pp.261-274.

Rocha, J.M. (1999), 'As Rotas de Imigração para Portugal', *O Público*, 14 May.

Rocha-Trindade, M.B. (ed) (1995), *Sociologia das Migrações*, Universidade Aberta, Lisboa.

Rowland, R. (1992), 'La Migracion a Grandes Distancias y sus Contextos: Portugal y Brasil', *Estudios Migratorios Latinoamericanos*, vol.7, no.21, pp.225-74.

Syrett, S. (1997), 'Cavaco Silva, The European Union and Regional Inequality (Regional

Development Policy in Portugal: 1985-95)', *International Journal of Iberian Studies*, vol.10, no.2, pp.98-108.

Veiga, U.M. (1998), 'Immigrants in the Spanish Labour Market', *South European Society and Politics*, vol.3, no.3, pp.105-128.

Williams, A.M. (1992), 'The Portuguese Economy in Transition', *Journal of the ACIS*, vol.5, no.2, pp.30-39.

Williams, A.M., King, R. and Warnes, T. (1997), 'A Place in the Sun: International Retirement Migration From Northern to Southern Europe', *European Urban and Regional Studies*, vol.4, no.2, pp.115-134.

Williams, A.M. and Patterson, G. (1998), 'An Empire Lost but a Province Gained: A Cohort Analysis of British International Retirement in the Algarve', *International Journal of Population Geography*, vol.4, no.2, pp.135-155.

6 The Liberalisation of Media and Communications in Portugal

HELENA SOUSA

A fundamental aspect of the rapid socio-economic change which Portugal witnessed in the post revolutionary period has been in the area of media and communications. Following dramatic changes in the 1974-75 revolutionary period, a second phase of comprehensive restructuring was undertaken by the Social Democrats (*Partido Social Democrata* – PSD) during their period of absolute majority rule between 1987 and 1995. The changes of this period are significant not only because of their importance to Portuguese economic and social development, but also because they provide insights into the political process of modernisation and liberalisation and the forces that drove this process forward.

Given the nature of political, economic and technological developments, changes in the media and telecommunications in this period were inescapable. Internationally the European Union (EU) was actively developing policies in this area whilst conservative governments in the UK, Germany and France, persuasively argued for the liberalisation of markets and the privatisation of state property. Important technological advances in the area of satellite, optic fibre and distribution technologies were also fundamentally changing the nature of the sector.

Nationally, significant changes were also taking place. The economy was growing rapidly and the liberalisation of the media and telecommunications markets was seen as inevitable. The pro-business approach of the Cavaco Silva led PSD governments favoured the privatisation of state media and telecommunications companies. Under the Social Democrats majority governments, newspapers which had been nationalised during the revolutionary period, returned to private hands, the radio sector was liberalised and one public radio station was privatised, whilst the television public service operator lost its monopoly and two national commercial television companies commenced operation. Similarly, and in accordance with EU policies, the telecommunications market was opened up to new actors and the three public telecommunications operators were merged and subsequently privatised. The Socialist governments led by António Guterres which followed inherited highly reformed but poorly regulated media and telecommunications sectors. To date these governments have made no structural changes, and instead have focused on attempts to strengthen existing regulatory bodies and improve a number of legal instruments in line with EU policies.

This chapter analyses the historical development of the Portuguese media and telecommunications sectors, focusing on three periods: the immediate post-revolutionary period and its aftermath (1974-87); pressures for change in the 1980s leading to the period of Social Democratic reform (1987-95); and the actions of the first Guterres government (1995-99). In seeking to understand the structural shifts which have taken place in both the media and telecommunications sectors, the analysis will look at the internal and external factors which have shaped change, and the manner in which the political process dealt with changing economic and technological realities.

The Media Sector, 1974-87

After the 1974 revolution, the media suffered a period of intense crisis. Pre-censorship was immediately abolished and swiftly followed by fierce confrontation over control of the most important media. The differences between the factions which comprised the so-called 'winners' of the revolution ensured that no consensus was possible over what role the media should play in a post-dictatorial society and a chaotic situation ensued. Mesquita et al. (1994) identify three conflicting tendencies in this period. First the inheritors of the old regime who tried to postpone the dismantling of censorship mechanisms. In terms of further legislation, this faction was in favour of *a posteriori* repressive measures.[1] Second the defenders of revolutionary Marxist ideas who were also favourable to *a posteriori* censorship. Third the adherents of a pluralist concept of the media, based on a Western model of democracy. This faction argued for the abolition of any form of censorship mechanisms, pointing out that the courts were the appropriate means for settling media disputes.

Notwithstanding these conflicting views, two major pieces of legislation approved after 1974 marked a move away from the past control of the media. The 1975 Press Law guaranteed that 'press freedom will be exercised without subordination to any form of censorship' (Art. 4). The 1976 Constitution similarly suggested that the pluralist view of the media had won through. It stated that the freedom of the press was guaranteed and no group was permitted to exercise censorship or obstruct journalistic creativity (Art. 39). Although these pieces of legislation were pluralist, in the sense that they expressed the view that different interests in society should have the right to express themselves and to influence the political process, their implementation exhibited a different conception. Due to the continued existence of dangerous 'reactionary forces', leftist elements within the *Movimento das Forças Armadas* (MFA) contended that the media had to be controlled during the revolutionary period. As a result, there was a clear contradiction in the MFA programme which contemplated both the 'abolishment of censorship and previous examination' and the creation of an 'ad hoc committee to control the press, radio, television, theatre and cinema' in order to 'safeguard military secrets and to prevent disturbances which could be provoked in public opinion by ideological aggressions from the most reactionary sections of society' (quoted in Bruneau & MacLeod, 1986, pp.165-66).

The battle for media control immediately after the revolution, and particularly after the 28th of September, was not only fought within the ad hoc Committee (which had powers to suspend and punish newspapers which did not conform to the leftist 'revolutionary' line) but also through the appointment of elements close to the MFA movement to leading posts in radio and television. By early 1975, the situation in the electronic media was chaotic; as Salgado Zenha's stated 'what is now going on in the *Emissora Nacional* and on television is very grave because there is not one censorship but several'(quoted in Mesquita, 1988, p.102). This highly volatile situation was further worsened by the installation of the communist provisional governments of Vasco Gonçalves after the 11th of March coup. The publication of the pluralist Press Law a month earlier did not prevent increasing levels of media intrusion.[2]

In this revolutionary period the press, which was still in private hands, was 'transferred' to public ownership. Three days after the leftist coup of 15th March 1975, important sectors of the economy such as banking and insurance, were nationalised. As many leading newspapers were owned by major economic groups and banks, they became state property. Of the important dailies only the *República* in Lisbon and *O Primeiro de Janeiro*, in Oporto remained in private hands (Mesquita et al, 1994). The nationalisation of the press was never explained as a political option rather it 'was presented as an indirect consequence of the nationalisation of the banking sector' (Mesquita et al, 1994, p.368). However, the reality was a desire to expand the scope of government's direct influence. Significantly, the nationalisation process was not reversed with the removal of the communist prime minister, Vasco Gonçalves, in November 1975.

Although under *Gonçalvismo* the electronic media were directly controlled by leftist forces, the more 'moderate' sixth provisional government increased the state's media ownership. Radio was nationalised, with the exception of the Catholic *Rádio Renascença* (RR) which had been in communist's hands under the governments of Vasco Gonçalves. The newly created national radio company, *Empresa Pública de Rádiodifusão* (EPR) later re-named as *Rádiodifusão Portuguesa* (RDP), created a radio duopoly with RR which remained untouched until the explosion of illegal radio stations in the mid-1990s. The television company RTP (*Rádiotelevisão Portuguesa*, SARL), which had been managed directly by the government after the coup (law-decree no. 278/74), was also nationalised in late 1975 (law-decree no. 674-D/75 of 02.12.75) and became a public company *RTP – Rãdiotelevisão Portuguesa, EP*. The RTP monopoly only ended with the opening up of TV channels to private initiative in the early 1990s.

Both RDP and RTP remained under the control of successive governments: 'since 1974 the eleven seats on the board of governors and the 20 directors posts at RTP and RDP have been held by 80 and 130 different people respectively, whose qualifications for the job were considered less important than their party membership cards'(Optenhogel, 1986, p.243). Indeed, administrations changed even more frequently than governments. The height of this instability was reached during the three and a half years of the *Aliança Democrática* (AD) which blatantly attempted to put radio and television at the service of the government. In the words of Victor Cunha Rego, the first chairman of the board appointed by the AD and President of the RTP from February 1980 to July 1980: 'impartiality in state television was unthinkable'(quoted in Bruneau, 1986, p.173).

What is particularly remarkable about the development of the media in Portugal is that laws drawn up during the exceptional revolutionary period continued to shape the media until the 1980s. This suggests that the authoritarian nature of the provisional leftist governments rather suited the newly created democrats. Despite the 1976 Constitution (with its impressive display of civil liberties) no elected government was prepared to grant freedom to the press. Following the political measures introduced during the revolutionary period, politicians from all affiliations consistently failed to redesign media policies. Instead they preferred merely to take the necessary steps to ensure that the nationalised media was favourable to them whilst in power.

The Telecommunications Sector, 1974-87

Although telecommunications are fundamental to economic development and state security, the Salazar regime paid little attention to civilian communications and maintained the basic market structure from the 1920s up until the 1960s. Marcello Caetano, on the contrary, perceived telecommunications as a means to help revive the economy and between 1968 and 1974 oversaw important investments in both internal and international telecommunications. However, despite these investments, Portuguese telecommunications were still backward when compared with most Western countries.

In the period from the 1974 revolution until the early 1980s no government introduced any significant change in the telecommunications sector. In 1981, a telecommunications reform was attempted and a Law Decree (188/81) was passed in accordance with the government's programme (Assembleia da República, 1981). This legislation recognised that the government had been unable to co-ordinate public communications operators due to a lack of infrastructures and contemplated the setting up of the *Instituto das Comunicações de Portugal* (ICP) to support the *Ministério das Obras Públicas, Transportes e Comunicações* (MOPTC) in the co-ordination of the telecommunications sector. Public postal and telecommunications services were to be maintained as a state monopoly, but the terminal equipment market was to be opened up to competition. However, the legislation was never implemented and the ICP was only eventually established in 1989; a reflection of political instability and a lack of commitment to the ICP by the 1983-85 socialist led government.

At this time it was widely accepted that Portuguese telecommunications were lagging behind the core EC countries and that digitalisation and optic fibre would have to be introduced if the country was to catch up. If agreement existed about the need to modernise and update the sector, no consensus existed on the strategy to be followed. Whilst some argued for a progressive introduction of new technologies so that Portuguese industry could adapt, political power favoured rapid change. The prevailing view was that 'only big multinationals could supply public digital exchanges given that there was no internal industrial or technological basis to do it'.[3] The 1983-85 government decided therefore to open up the switch contract bidding to national and international companies not only for economic reasons but also to 'demonstrate openness to the European Economic Community' (Noam, 1992, p.261).

In addition to the procurement contracts for digital exchanges, 1985 was also important for national telecommunications as the first optic fibre cable was installed, allowing the future introduction of new services such as video conferencing and cable TV (Santos, 1989). This was followed by the installation of the first Siemens' digital exchange in June 1987 by TLP (TLP, 1992). The mid-1980s was therefore a turning point in terms of network modernisation yet in a manner that failed to safeguard Portuguese industrial capability.[4] Rapid technological change was perceived to be of the foremost importance on the eve of Portugal's entry into the EC, even if this was at the expense of the national telecommunications industry.

Pressures for Change

Given the nature of political, economic and technological developments in the mid-1980s, changes in the media and communications became inevitable. At the European level, the EU was developing policies for television and telecommunications. Internationally, conservative governments in the UK, Germany, France and the US propounded the case for the liberalisation of markets and privatisation of state property. Furthermore, important technological advances - particularly the development of satellite and optic fibre and the subsequent convergence of distribution technologies - had enormous implications for the sector. With the proliferation of European satellite TV channels critics of the national RTP monopoly within Portugal began to argue that once international private TV channels could be received, there was no reason why national private channels should not also exist.

Important national level changes were also taking place. Up to the mid-1980s, the political instability in the country was so acute that any comprehensive set of political decisions was hard, if not impossible, to implement. In 1987, one year after Portugal joined the European Community (EC), the first majority government since the 1974 revolution was elected. At this time the country's economy was booming. One consequence was a substantial rise in total advertising revenue from around £52 million in 1986 to £400 million in 1994.[5] In this economic context relatively independent newspapers such as *O Independente* and *Público* were set up, seriously reducing the government's ability to suppress politically damaging material. In addition, the climate of opinion turned against the concentration of the media in the state's hands. The Cavaco Silva government itself believed that if Portugal was to be seen as a truly European partner, changes in the economy, and consequently in the media market, had to be introduced. A pro-business approach was taken and the liberalisation of the media market and privatisation of a substantial share of state media became imminent.

In certain respects the subsequent attempts of the Cavaco Silva administration to open up the media and pick winners echoed similar attempts by Marcello Caetano in the 1970s. Caetano tried to maintain power over the press by allowing economic groups close to the regime to own periodicals. At a time when the country was supposedly opening up and censorship appeared set to be abolished, Caetano urged economic groups to buy up newspapers. In the different context of the late 1980s and early 1990s, Cavaco Silva also carefully chose the actors who would be allowed to participate in the newly liberalised broadcasting market and privatised state press. For the government, if liberalisation and privatisation could not be avoided, the media had to be in the safest possible hands.

In the telecommunications sector, external factors were particularly important in driving forward change. By 1986, the EC's telecommunications policy was in existence, designed around two contradictory trends: the support of the most competitive information technology (IT) companies through research and development (R&D) programmes; and following the US example, the endorsement of more competition and liberalisation of equipment and services. Following Portugal's accession to the EC, the pressure to modernise and reorganise the telecommunications sector intensified. In May 1986, Sequeira Braga, head of *Secretaria de Estado dos Transportes e Comunicações* (SETC), commissioned a study to examine Portuguese telecommunications and to report on the most pressing issues in this area. The resulting study of the *Comissão para o Estudo do Desenvolvimento Institucional e Tecnológico das Comunicações* (CEDITC) presented in May 1987 (MOPTC, 1987) concluded that Portuguese telecommunications were lagging behind those of most European countries by an estimated 15 years. To remedy this situation it recommended that:

- The Institute of Communications (ICP) should be created and regulatory functions should be transferred to this institute;
- A holding company should be set up to increase the efficiency of

operators (*Correios e Telecomunicações de Portugal* (CTT), *Telefones de Lisboa e Porto* (TLP) and Marconi). This company should pursue a global and integrated strategy for the national telecommunications sector;

- Postal and telecommunications services should be separated. Until then, CTT covered both activities;
- Measures should be taken to prepare the opening up of telecommunications operators to private capital;
- More flexibility should be introduced into new telecommunications services. Possibly, competition should be introduced;
- Prices should be liberalised in the non-basic telecommunications sub-sector.

These recommendations - which were broadly in line with what was being discussed at the European level at the time - led to a dramatic increase in political activity surrounding telecommunications in Portugal. The recommendations were largely adopted by the 11th Constitutional Government (Assembleia da República, 1987) and constituted an important input to the 1989 Basic Telecommunications Law; a law notable in the legal history of Portuguese telecommunications because the general principles of the sector were compiled in the same piece of legislation for the first time.

Opening up the Media Market, 1987-95

Between 1987-95, the two PSD majority governments undertook the most comprehensive changes in the media since 1974-75. Although the published governmental programmes (Assembleia da República, 1987 and 1992) did not clearly set out the government's objectives for the sector, the following lines of action were evident across these two programmes:

- The nationalised press to return to the private sector
- A minimum radio and television public service to be provided by the state
- The radio sector to be liberalised and/or privatised (e.g. Rádio Comercial)
- A television act to be approved in order that two TV channels could be granted to private operators
- The national news agency, *Lusa*, to continue to be publicly owned
- Attention to be given to the Portuguese communities abroad, to Portuguese speaking peoples, and to the regional press and professional training.

Despite the liberalisation agenda, certain of these lines of action were directly related to the continued state control of content. The state's ownership and subsequent government control of the national news agency (*Lusa*) for example, was crucial for the executive's dominion over political content both in the national and local media. This was mainly due to the lack of human and material resources in the media which forced them to rely heavily on *Lusa*.

Other policy proposals dealt with the structure of the media and consequently marked an important shift from previous minority governments. The first set of measures directly related to the structure of the media concerned the reorganisation of the radio broadcasting sector. As early as 1976 there had been calls to legalise local and regional radio stations but no government had been willing to undertake this. By the mid-1980s there were so many illegal radio stations operating that the government could no longer ignore them. Nevertheless, it was only in 1989 that 310 local frequencies were finally allocated. In the following year, two regional frequencies were attributed: one to *Rádio Press*, part of the Lusomundo group, and the other to *Correio da Manhã Rádio*, which belonged to the Carlos Barbosa group.

These actions to liberalise radio media were followed, in 1991, by the privatisation of the two most important state owned and controlled newspapers. In the context of the government's wider privatisation programme the maintenance of *Jornal de Notícias* and *Diário de Notícias* under state control was unjustifiable. The government was caught in a dilemma between the perceived need to control these newspapers and their ideological and political belief in privatisation. In a controversial process both papers were bought by *Lusomundo*, one of the most important multimedia groups in Portugal, and one that was perceived (at the time) to have close links with the government.

By far the most important aspect of Cavaco Silva's governments' media policy was the opening up of TV channels to private ownership. There had been considerable debate about private television in Portugal following the 1982 Constitutional changes. The 1976 Constitution stated that no TV channel could be privately owned and the 1982 review did not contemplate any change in this respect. Nevertheless the review process did re-open controversy on the issue. The Catholic Church was one of the first actors to openly express its desire to own and run a private TV channel and from 1987 onwards, the Balsemão, Sonae and Presslivre economic groups started to seriously evaluate this possibility given the new context of economic growth and political stability.

The process towards the creation of private television was initially delayed because the government concentrated its efforts on the reorganisation of the radio sector. The constitutional obstacles were removed on the 1st of June 1989 when the National Assembly approved amendments to the legislative text to permit TV channels to be privately owned. The next highly controversial step was the drafting of a new television act. A number of conflicting interests were evident and the Catholic Church was at the centre of the polemic. The Church wanted to be granted a TV channel without participating in the bidding process. So, when the government laid out its proposals the Portuguese bishops argued publicly against them stating that 'the proposed law does not correspond to former commitments and to what was expected; it does not safeguard the Church's rights consigned in the Constitution'(*Público*, 1992b, p.5).[6] In the middle of this fierce debate Parliament approved a new television law on the 13th of July 1990 which did not provide any privileged position for the Catholic Church, but did not prevent the Church from applying for a channel.[7]

Once the new television law was passed and the bidding regulations approved, on the 2nd April 1991 three candidates applied for the two available national TV channels: the *Sociedade Independente de Comunicação* (SIC) led by Pinto Balsemão; TV1 *Rede Independente*, chaired by Proença de Carvalho with the support of the Carlos Barbosa media group (Presslivre), and *Televisão Independente* (TVI), close to the Catholic Church. The three candidates put forward quite different projects which the government, with the approval of the *Alta Autoridade para a Comunicação Social* (AACS – High Authority for the Media), had to choose between. The Proença de Carvalho candidacy (TV1) promised generalist and popular programming comprising news programmes, *telenovelas*, talk-shows, movies, series and sports. The emphasis here was on national production and the exclusivity of national capital. The slogan of SIC led by Pinto Balsemão was that of 'difference, popularity and intelligence', but with an emphasis on information, with four proposed news bulletins per day and a commitment to change programming when the news justified it. SIC was also to include series, talk-shows, competitions and movies. The Catholic Church's candidacy (TVI) proposed a channel of 'Christian inspiration' of true 'quality' and 'public utility'. Programming was based around three news bulletins a day and, like the other projects, a mixture of series, movies, sports, quizzes and talk-shows. Although not a religious channel, TVI put forward in its candidacy space for religious content.

From these candidates the Government and the AACS had to make a decision. Politicised and without resources and credibility, the AACS was not prepared to put forward its views on such a sensitive issue. Yet, as its opinion was required by the Constitution, the AACS made a decision on the basis of 'technical equality' which meant no candidacy was excluded. The AACS judgement was that the TV1 project (Proença de Carvalho) was 'deliberately ambitious', TVI's (Church) 'modest', and SIC's (Balsemão) 'balanced'. As a result it was exclusively up to the government to take the final decision on the issue. Although in the beginning other senior politicians were involved, when final decisions were taken, Cavaco Silva managed the process himself.

On the 6th of February 1992, after a Cabinet meeting, *ministro* Marques Mendes, publicly announced the results: SIC was attributed the third and TVI the fourth, national channel. According to Marques Mendes these decisions were taken after considering the opinions of the AACS alongside four additional criteria: technical quality, economic viability, type and characteristics of the programming, and the candidates ability to satisfy diversity and public interest (Presidência do Conselho de Ministros, 1992). For the opposition and for TV1, the result was little more than a 'political decision' lacking in any transparency. The editorial of *Público* newspaper stated: 'The government took the less politically damaging decision attributing the two private channels to the Church and to Balsemão, the candidates with more "specific weight"' (*Público*, 1992b).

At this time attention was almost exclusively concentrated on who would gain control over the two new TV channels. This is hardly surprising given that, until then, the state which owned RTP had effective editorial control over the company's output. If the same were to happen with the new TV stations, politicians holding office would have to be extremely careful as to who 'deserved' such a powerful instrument. In consequence all other crucial issues associated with the opening up of the television market - such as sources of financing, balanced programming and national production - were neglected. The government thus decided to attribute two licenses and abolish the television license fee without ever examining whether the broadcasting market could support four national television channels. The resulting Television Law (58/90 of 7 September) was so badly drafted that it was totally ineffective, allowing TV channels to take the easy option of cheap imports and populist programmes.

Opening up the Telecommunications Market, 1987-1995

In line with what was being discussed in the EC at the time and following the *Comissão para o Estudo do Desenvolvimento Institucional e Tecnológico das Comunicações* (CEDITC) recommendations, the XI Constitutional Government decided to introduce major reforms to the telecommunications sector. Of the various CEDITC recommendations, the first to be implemented was the separation of the regulatory and operational functions (until then the CTT/TLP exercised both functions). The EU and other pro-competition international actors had argued that the dual regulatory and commercial function of the telecommunications operator could not be sustained in the new competitive environment because of a conflict of interests.

Law decree 283/89 of 23 August 1989 brought the ICP legally into existence and was intended to create an environment which would allow an even-handed introduction of competition into telecommunications services in line with EC directives.[8,9] Yet this legislation failed to grant real autonomy to the regulatory institute. Despite the ICP's array of responsibilities,[10] with the exception of technical matters, its role was merely advisory and supportive. Importantly, no line was clearly drawn between policy and regulatory issues. Furthermore, the members of the board of directors were to be appointed by a resolution of the Council of Ministers, so even with its legally recognised administrative and financial autonomy,[11] the ICP could not be said to be a truly independent body.

Immediately following the legal setting up of the ICP, the Basic Law on the establishment, management and exploitation of telecommunications infrastructures and services (88/89 of 11 September 1989) was approved by the National Assembly. This Telecommunications Act marked a watershed in Portuguese telecommunications legal history because, for the first time, the general principles for the sector were compiled in the same piece of legislation. Under this law, apart from fundamental services, which would continue to be provided by the state, other actors - either private or public - could now apply to become services providers. Only the telecommunications infrastructure was to remain firmly under the responsibility of the public telecommunications carrier (article 7). Although new entrants were required to comply with specific rules and regulations, the Telecommunications Act also set out requirements concerning protection of competition.[12] Unfortunately these legal provisions did not prevent alleged abuses of dominant positions and accusations of unfair competition.

The introduction of these legal instruments was not particularly controversial as both the government and the main opposition party, the Socialist Party, perceived change as necessary and inevitable. The creation of the ICP and the liberalisation of complementary and value-added services were directly and indirectly related to the EU's legal framework. If Portugal had not passed this legislation it would have had to comply later on with the ONP Council directive (90/387/EEC), with the Commission's Services directive (90/388/EEC), and subsequent legislation.[13] In the Portuguese case the liberalisation was nevertheless quite limited given that before the partial privatisation of *Telecom Portugal* only 3% of the telecommunications market was in the private sector (*Diário de Notícias*, 1994).

Although these changes were ultimately introduced by the national parliament and government, the role of the EU was crucial. As previously discussed, the EU persuasively convinced the member states that no alternative existed but to open up their markets. Core countries had very concrete interests in doing so whilst peripheral countries were convinced that they had not much to lose. In consequence the EU's Council of Ministers approved legislation that made it more difficult for member-states to take protectionist measures. For Portugal, the liberalisation process was therefore seen as inevitable once no other alternatives existed.[14]

In addition to the opening up of the telecommunications market to new entrants, the Portuguese authorities believed that the three traditional public operators (CTT, TLP and Marconi) required reorganisation.[15] These operators were, for historical reasons, organised on a geographical basiswhich was considered by the government as inappropriate.[16] In 1992 a financial holding society, *Comunicações Nacionais* (CN), was set up (Law decree 88/92 of 14 April 1992) mainly to co-ordinate the sector, to define investment/business strategies and to deal with the privatisation process. CN started its operations in early 1993 and comprised five independent public companies: the former postal services of *Correios e Telecomunicações de Portugal* (CTT), *Telecom Portugal* (TP) (the telecommunications arm of the earlier CTT),[17] *Teledifusora de Portugal* (TDP),[18] *Telefones de Lisboa e Porto* (TLP) and *Marconi*.

The creation of CN was justified by its chairperson, Cabral da Fonseca,[19] as the 'rational' way forward for the sector (*Público*, 1992c). This view, however, was soon challenged as Telecom Portugal put in place its strategy to become the dominant actor in the Portuguese telecommunications scene. The president of Telecom Portugal from 1990 to 1992, Gonçalo Areia, publicly argued for the setting up of a single telecommunications operator (*Expresso*, 1992) but it was the subsequent president, Luís Todo Bom, (also vice-president of the party in power) who convinced the CN and the government that TP was the only company capable of leading the reorganisation process.

Under this plan TP was to take over TDP, TLP and Marconi. Despite Marconi's opposition and fierce criticism from the government's own ranks, the merger went ahead. The CN's President, Cabral da Fonseca, in a notable change of position, argued that for more than one company to provide basic services in a country with ten million inhabitants and current levels of economic development, was unthinkable: 'there is no way of surviving in a competitive environment' (*Expresso*, 1993). Interestingly the reasons why a single telecommunications operator would provide a better service than had traditionally been provided by the three existing operators was never clearly laid out. Luís Todo Bom argued in very general terms that Portugal should follow the Dutch model, and that the creation of a single telecommunications operation was essential to fight foreign competition in anticipation of full market liberalisation between 1998 and 2003 (*Público*, 1993). However, neither the government nor TP were able to justify the paradox of arguing simultaneously for liberalisation and concentration. On the one hand, it was argued that liberalisation and full competition could only benefit the consumer; on the other hand, that only a big operator could respond to the challenges imposed by liberalisation. In any case, because the reorganisation process was designed behind closed doors with no public debate and adequate consultation with long standing actors in the field, there was no need to prepare a consistent case. The process was conducted with speed but lacked participation.

The partial privatisation of Telecom Portugal cannot be seen as a direct result of EU policies. According to O Siochrú (1993), the only fear that the EU expressed for the 'less favoured regions' was their ability to find the necessary resources to keep pace with liberalisation. But, even if the EU had tried to convince its members to privatise, it had no open policy on the issue and member states were entirely free to decide their own strategies. Portugal therefore made a conscious choice to join the North American/European privatisation bandwagon. Government thinking on this issue was exemplified by *ministro* Ferreira do Amaral who publicly stated that - whether we want it or not - the telecommunications sector will be exclusively private because public companies have no agility nor vocation to stand a chance in a competitive market:

This is happening in all countries in the world. I do not know any [country] which is, at this stage, thinking about nationalising the telecommunications sector and the vast majority are thinking about privatising.[20]

In this speech Ferreira de Amaral illustrated two predominant lines of thought: if most countries are privatising, Portugal must do it as well; even if Portugal resists privatisation, it would happen anyway. Yet, Ferreira do Amaral did not attempt to explain the benefits of privatisation and why public companies (whose managers have been appointed by the social democrats over the last decade) had performed poorly. The telecommunications sector in Portugal was in private hands from the 19th century up until the 1960s, and in the public sector ever since. In both periods it performed deficiently.

On the 20th of March 1995, just before privatisation went ahead, the government granted TP a public service concession contract for 30 years (due for renewal in 15 years). According to this contract,[21] TP gained the exclusive right to provide basic telecommunications services and to manage all telecommunications infrastructures which support these services although the infrastructures would remain in the public domain.[22] The company therefore had a monopoly on basic services until this area was liberalised in January 2000. Following the first stage of TP's privatisation the Council of Ministers approved the abolition of *Comunicações Nacionais*. CN had co-ordinated the restructuring of the sector and prepared the privatisation of TP, but had been unable to lead the process and was too weak to fight lobbies and balance the needs and interests of the various actors involved in national telecommunications. Ferreira do Amaral stated that the CN would come to an end because it has accomplished its mission, whilst Cabral da Fonseca said he would now assist the evolution of the sector because 'the essential had already been done and it is irreversible' (*Público*, 1995, p.32).

Guterres Government: Tuning up Reforms

The incoming Socialist government of 1995 inherited a transformed media and communications system, but one that was characterised by poor regulation and weak regulatory bodies. António Guterres' government programme (Assembleia da República, 1995) did not contemplate any significant changes in the media/communications system instead proposing to strengthen existing regulatory bodies and to improve legal tools.

The government's programme referred to communications in three different dimensions; mass media, telecommunications and information society. For the mass media, the Guterres government priorities included: the right of information; revitalisation of the media sector; independence in the management of the media public sector; and the media as a tool of international politics. In the telecommunications sphere the programme aimed to promote measures which would guarantee real competition in telecommunications services, and introduce liberalisation in line with EU directives and market changes. Regarding the so-called 'information society' the government identified a crucial role for IT in the development of 21st century society and consequently that new technologies should be used to transform social and economic life.

Though it is still too early to analyse comprehensively the impact of the policies of the first Guterres administration (1995-99), it is possible to examine some aspects of the implementation of their programme. In the broadcasting arena, a new Television Act was passed (Law 31-A/98 of 14th of July). This new Act introduced changes in both the access to and exercise of television activity and, for the first time, created the legal possibility to establish local, regional and thematic channels.[23] Although cable and satellite television were well established, companies had not previously been allowed to produce their own programmes (Sousa, 1999). The new television law therefore opened the floodgates for thematic channels. In response terrestrial television companies sought to associate themselves with cable operators and international content producers in order to guarantee their places in this new broadcasting context. SIC, for example, linked up with the Brazilian network *TV Globo* and the biggest national cable operator, *TV Cabo* in 1998, in order to develop the *Premium TV* project which offers two codified movie channels (Telecine1 and Telecine2). Similarly in 1998 RTP signed a contract with *TV Cabo* and a company with multiple interests in sports, *Olivedesportos*, to create a consortium that offers a codified Sports channel, *Sport TV* (Sousa, 1999). However, according to the *Secretário de Estado da Comunicação Social*, Arons de Carvalho, local/regional television channels will take longer to develop and are not expected to be licensed before 2005 or 2006 (Carvalho, 1999).

The proliferation of television channels does not necessarily mean that the financial situation of broadcasting companies improved during the Guterres' administration. In fact, TV stations such as RTP and TVI have experienced important financial losses in recent years. The advertising market is small and, apart from SIC, terrestrial broadcasting companies have suffered from highly unstable management, a lack of advertising revenues, and accumulated debt. When the broadcasting market was opened up to private initiative in 1992, the Cavaco Silva government abolished the television license-fee and sold RTP's transmission network to Portugal Telecom. These political decisions placed RTP in a difficult economic situation and transformed so-called Public Service Broadcasting into standard commercial television (i.e. RTP had to fight for audiences in order to obtain a significant slice of the advertising cake). As a result, RTP was neither performing its duties as a public service nor operating as a successful commercial company. In response, the Guterres' government chose not to reverse the abolition of the license fee, but instead attempted to modify RTP's ambiguous position through the creation, in December 1996, of a new Public Service Contract between the State and the public operator. Under this new contract Public Service is seen as a mission and a programming philosophy. Whilst the contract defines more clearly RTP's objectives, it has not solved its financial problems.

Aside from the new television law, the government has also promoted the introduction of digital television. The prime minister announced in August 1998, that this would be introduced 'as soon as possible', although in fact terrestrial digital television was not expected to be introduced before 2001. In preparation in 1998 the *Instituto da Comunicação Social* (ICS) and the Portuguese Communications Institute (ICP) co-ordinated a public consultation process on Terrestrial Digital Video Broadcasting (DVB-T).

Not unlike its Social Democrat predecessors, the Socialist government also considered international broadcasts to be of great importance to its foreign policy. Under the PSD's tutelage, RTP's International channel (*RTP Internacional*) and RDP's International channel (*RDP Internacional*) were launched. Recognising the relevance of these, Guterres supported their expansion and consolidation. Most Portuguese communities abroad and Portuguese speaking nations now have access to these international television and radio channels. In addition, the Guterres government decided, in collaboration with RTP and RDP, to set up channels specifically designed for Portuguese speaking African countries. *RTP África* and *RDP África* have indeed established themselves as news and programming sources in Lusophone Africa, and the Lusa news agency has also reinforced its links with Portuguese speaking African countries.

In radio broadcasting, the socialist government also developed and/or revised a number of legal tools. It made it compulsory for local radios to produce their own content (most were simply broadcasting national radio stations feed) and created financial incentives to achieve this, namely subsidising technological modernisation, providing institutional advertising, and reducing telecommunications costs (via an agreement with TP).

In the government's published programme, the independence of the public sector media and journalists' rights were high on the agenda. Both the Press Law and the Journalists' Statute were revised with the intention of expanding pluralism and independence within media companies and reinforcing journalists' rights. Changes in the media regulatory body (AACS) were also introduced. The first Guterres administration altered its composition and widened its powers in response to the perception that this body was highly politicised and had limited influence. To date, most of these changes have yet to be proved effective. The shadow of state control still hangs over both RTP and RDP and periodical allegations of interference continue to be made by journalists.

In the telecommunications sector, Law-decree 381-A/97 of the 30th of December 1997 was particularly significant. This law, itself a transposition of a series of EU regulations to the national legislative body (Sousa, 1999),[24] introduced the principle of 'freedom of establishment' with the aim to reduce bureaucracy and permit easier entrance of new actors into the market. The law established a new telecommunications access regime for a public telecommunications networks operator and provider of public telecommunications services, and aimed to simplify access to the telecommunications market. In consequence a number of telecommunications services no longer require authorisation from the regulator.

The effects of the ongoing liberalisation of the telecommunication market is particularly notable in the rapidly expanding mobile phone sector. Growth in this sector was further stimulated by the entry of a new mobile phone operator, *Optimus*, in 1998. *Optimus*, a *Sonae* group project, forced other operators (Telecomunicações Móveis, TMN and *Telecel*) to reduce tariffs as a result of an aggressive pricing policy which saw it achieve half a million 'pre-adherents' by the time of its launch. Despite substantial investments (of between 350-500 million Euros in the first three years), the *Sonae* holding company expected *Optimus* to break-even in 2001.

In an attempt to develop and implement policies to promote an 'Information Society' the Portuguese government published the 'Information Society Green Paper' in 1997. Increased use of information and communication technologies meant that by 1998, 20% of the Portuguese population had access to the Internet, although only 10.6% used it regularly. Studies showed that the Internet was used mostly at school, then at work and lastly at home, and was most important in the education (amongst academics and students) and service sectors (e.g. banking, insurance, advertising and travel agencies) (Sousa, 1999). In response to this context the Green Paper spawned a number of initiatives which aimed to improve the use of, and access to, new technologies. In the arena of R&D, a 'National Science, Technology and Society Network' was created, a scientific network which aims to bring together national researchers and stimulate and consolidate R&D activity. Policies to encourage widespread use of new technologies also include attempts by the *Ministério da Ciência e da Tecnológia* (Science Ministry) to introduce the Internet into every school in the country (from the 5th to the 12th grade) and to extend its use in universities, libraries, and research centres. A further project, 'Computers for All', had the objective of increasing the number and usage of Internet connected computers at home. In addition, a considerable number of small-scale initiatives, such as the creation of tele-work centres, were also under way.

Conclusions

This chapter has placed the fundamental changes that have occurred in the media and telecommunications sectors in the post-revolutionary period within their historical context. Analysis of the recent evolution of these sectors richly illustrates the politically controlled nature of the modernisation process in Portugal. As key sectors, not only economically but also politically and socially, the preceding analysis has drawn out the nature and types of change and illustrated how these processes have been driven by a mixture of political self-interest, ideology, and economic and political necessity.

The Cavaco Silva majority governments oversaw structural changes in all aspects of the media and communications; from telecommunications to the press all media and communications sub-sectors were transformed in this period. Due to the particular conjuncture of external and internal factors these changes came to be viewed as something of an 'inevitable' process. EU policies provided a context which strongly drove forward the modernisation process and liberalisation agenda, particularly within the telecommunications sector. However, the PSD government chose a particular road towards modernisation; one characterised by an ideological desire for greater economic liberalism and a political desire to retain state control. By the time António Guterres came to power in 1995, his administration had a much reduced room for manoeuvre. Major changes in market structure had already been introduced and were perceived to be largely irreversible. In this period the EU intervened more frequently and more consistently in the telecommunications arena, and thus between 1995-99 government action was largely driven by the need to conform to EU directives. In the media, where the EU remains far less influential, the first PS government did attempt to improve perceived deficiencies inherited from the previous government by strengthening regulatory bodies and legal tools.

Notwithstanding the differences between the Cavaco Silva and Guterres governments concerning intervention in communications, there are also some important differences regarding the mode of governance. During the Cavaco Silva period, political power was largely concentrated in the hands of the prime minister and a few senior politicians. Despite the formal dispersion of power (a number of *Secretarias de Estado* and government bodies were involved in policy-making), all fundamental aspects of communications/media policy were dealt with by the prime minister himself. Most decisions were taken behind closed doors with no justification provided. This lack of openness contributed to the absence of informed debate over the strategies pursued, particularly with regard to whether they were the best way to serve the national interest in terms of Portugal's economic capacity and cultural development. In contrast, the Guterres government tried from its earliest days to involve social actors in the decision-making process. 'Dialogue' has been a key word and has had some real meaning in the communications arena. Indeed, despite the continual formal fragmentation of power in this policy-area, the different *Secretarias de Estado* involved in the sector have attempted to implement a more open decision-making process, promoting public consultation and publicising the distinct stages of the political process.

Notes

1   As the regime collapsed with almost no resistance, former supporters of the regime were amongst the 'winners' of the revolution.
2   Vasco Gonçalves' disliked the extensive freedoms guaranteed to journalists by the Press Law and so in parallel created a new organism, the *Conselho de Informação*, to direct the state's media agenda in order to construct a 'People's democracy'.
3   Personal interview with J. Tribolet, chairperson of INESC (*Instituto de Engenharia de Sistemas e Computadores*), Lisbon, 17th November 1994.
4   Interview with J. Tribolet, 17th November, 1994.

5   These figures are estimates based on data from Sabatina and the opinions of several experts in the field.

6   In the 1970s, the Church was granted an assurance by the former prime minister, Sá Carneiro, that it would be attributed a television channel. Hence, religious leaders felt they were being unfairly treated. The Catholic Church's *Rádio Renascença* network was used to put forward these arguments and clerics throughout the country were given the task of reading and commenting on the bishops' position.

7   This law states that 'the activity of television cannot be exercised and financed by political parties or associations, unions, professional and employers organisations, and by local authorities'(law no. 58/90, Art.3). Significantly, religious organisations are not mentioned thus enabling the Catholic Church to enter the competition.

8   The ICP was formally created in 1981 under the law decree 181/81 of 2nd of July but this legislation was never enacted so the ICP did not come into operation until after the 1989 legislation.

9   The ICP was set up in anticipation of the EC directive on competition in the markets for telecommunications services (90/388/EEC of 28 June 1990).

10   According to article 4 of law decree 283/89 the ICP's responsibilities included: to actively contribute to the sector's legal framework; to provide assistance to the government for the purposes of carrying out its tutelage responsibilities; to co-ordinate, on a national level, all matters concerning the carrying out of treaties, conventions and international agreements; to approve materials and equipment, to undertake the management of the radioelectric spectrum; and to license public sector communications operators in addition to providers of value added services.

11   The ICP had to generate its own financial resources mainly from spectrum management (spectrum users paid ICP directly) and from the issuing of licenses.

12   For example public telecommunications carriers should guarantee the use of their networks for all communications carriers under equal competitive conditions. When public carriers provide complementary services they are equally forbidden to use any practice which may distort conditions of competitiveness or which are considered to be an abuse of a leading position (article 14).

13   Such EC Directives are indirectly binding in the sense that it is up to the member states to decide how the intended results of the legislation are to be achieved.

14   Speech of Ferreira do Amaral on the 5th APDC Congress in Lisbon, November 1994.

15   Marconi was considered a public operator but 49% of its shares were in private hands.

16   CTT used to supply local telephony to the entire country with the exception of the two main cities, Lisbon and Oporto, and long distance communications to Europe and North Africa. TLP used to cover Lisbon and Oporto while Marconi had the monopoly of cable and satellite communications to the rest of the world.

17   The separation of the CTT's post and telecommunications activities was set out in law decree 277/92 of 15 December 1992.

18   TDP was set up through law decree 138/91 of 8 April 1991 in order to distribute broadcasting signals for RTP and for the forthcoming television companies.

19   Cabral da Fonseca was perceived as an ally of Ferreira do Amaral and, before chairing CN, was the chief of the Portuguese commissioner's cabinet in Brussels.

20   Speech delivered at the APDC conference in Lisbon, November 1994.

21   See law decree 40/95 of 15 February.

22   What was privatised was the provision of the services, not the infrastructures.

23   Prior to this the Portuguese television broadcasting system comprised a number of channels: two public national channels (RTP1 and RTP2), two private national channels (SIC and TVI), two public regional channels (RTP-Açores and RTP-Madeira), and two public international channels (RTP África and RTP Internacional).

24   These EC directives comprise: 96/2/EC (mobile and personal communications); 96/19/EC (introduction of full competition in the telecommunications market); and 97/13/EC (common framework for authorisations and licenses in terms of access to the telecommunications market).

References

Assembleia da República (1981), *Programa do VII Governo Constitucional*, Apresentação e Debate, AR-Divisão de Edições, Lisbon.

Assembleia da República (1987), *Programa do XI Governo Constitucional*, Apresentação e Debate, AR-Divisão de Edições, Lisbon.

Assembleia da República (1992), *Programa do XII Governo Constitucional*, Apresentação e Debate, AR-Divisão de Edições, Lisbon.

Assembleia da República (1995), *Programa do XIII Governo Constitucional*, Apresentação e Debate, AR-Divisão de Edições, Lisbon.

Bruneau, T. and MacLeod, A. (1986), *Politics in Contemporary Portugal, Parties and the Consolidation of Democracy*, Lynne Rienner, Boulder.

Carvalho, A. A. (1999), Communication delivered at the *VII Congresso Nacional de Rádios*, 27-28 February, Óbidos, Portugal.

Diário de Notícias (1994), 19 November, p.12 (Negócios).

Expresso (1992), 19th December (Economia).

Expresso (1993), 20th November.

Mesquita, M. (1988), 'Estratégias Liberais e Dirigistas na Comunicação Social de 1974-1975, da Comissão Ad Hoc á Lei de Imprensa' in *Comunicação e Linguagens*, no.8, December, pp.85-113.

Mesquita, M. et al. (1994), 'Os Meios de Comunicação Social' in A. Reis (ed) *Portugal, 20 Anos de Democracia*, Círculo de Leitores, Lisbon, pp.360-405.

Ministério das Obras Públicas, Transportes e Comunicações (MOPTC) (1987), *Desenvolvimento Institucional e Tecnológico das Comunicações*, MOPTC, Lisbon.

Noam, E. (1992), *Telecommunications in Europe*, Oxford University Press, Oxford.

Optenhogel, U. (1986), 'Portugal' in H.J. Kleinsteuber et al. (eds) *Electronic Media and Politics in Western Europe*, Campus Verlag, Frankfurt, pp.239-250.

O Siochrú, S. (1993), 'The EC's Telecommunications Policy and Less Favoured Regions: the Role of the STAR Programme', paper delivered at the *IAMCR Conference*, Dublin, 24-26 June.

Presidência do Conselho de Ministros (1992), 'Atribuição dos Novos Canais de Televisão', *Press Release*, 6 February.

Público (1992a), 7th February, p.5.

Público (1992b), 7th February, p.3.

Público (1992c), 9th December.

Público (1993), 10 September, p.35.

Público (1995), 14th July, p.32.

Santos, R. (1989), 'Entre a Telefonista e a Central Digital - TLP: A História de uma Empresa Centenária', unpublished communication, Vila Nova de Gaia.

Sousa, H. (1999), 'Portugal' in *European Audiovisual Sector 1998: Regulation and Practice*, European Audiovisual Observatory, Council of Europe.

TLP (Telefones de Lisboa e Porto) (1992), *1982-1992, 110 Anos a Comunicar*, TLP, Lisbon.

7 Environmental Issues in Portugal: Towards a Sustainable Future?

CARLOS PEREIRA DA SILVA

Introduction

The idyllic notion of Portugal as 'a small garden planted by the seaside' emerged as a result of an industrial development process, which proceeded more slowly, and at a later date than elsewhere in Western Europe. This image, which is far removed from reality, has contributed towards a contemporary picture of Portugal as a country lacking environmental problems. However, industrial development since the 1960s combined with a lack of any overt concern for environmental issues, has resulted in severe impacts upon the environment. It has been these processes of change over the last thirty years, which are responsible for the negative environmental inheritance that exists in Portugal today.

To some extent this position is understandable. In a country with traditionally low levels of socio-economic development where a large part of the population lacked basic necessities, it was perhaps unreasonable to expect anxieties about, and awareness of, the environment to be a central concern. During the 1960s, the movement of the Portuguese population to the coast or abroad in search of a better quality of life exacerbated the division between the interior and the coastal margin of the country. The interior of Portugal became increasingly characterised by an ageing population and the widespread abandonment of traditional activities such as agriculture. The impact of this change upon the Portuguese environment was a major degradation of the traditional rural landscapes which had emerged through years of human-environment interaction. One high profile consequence of such change has been the dramatic rise in the number of forest fires as the result of the abandonment of traditional forest management practices. Another consequence has been the concentration of population into the coastal areas of Portugal. In a small country totalling only 90,000 square kilometres, more than 70% of the population is now concentrated into a western coastal fringe measuring no more than 20 kilometres wide. It is in this fringe that most of the country's environmental problems are concentrated. Industrial activities, which show little concern for the environment, are the principal source of current environmental problems and are condensed into this area, particularly within the two main metropolitan areas of Lisbon and Oporto.

Immediately after the revolution in 1974, the state's main socio-economic objectives were to meet the population's basic needs, increase salaries, and improve working and living conditions. As a result, even though approaches to dealing with environmental issues changed dramatically in the latter half of the 1970s, the financial investments necessary to create change remained minimal. Accession to the European Community (EC) provided the basis for significant qualitative and quantitative leaps in the way that environmental issues were perceived and tackled. After joining the EC in 1986 the environment began to receive more attention, as the need to transpose European environmental legislation into Portuguese law became apparent. Moreover, through European Union (EU) funding Portugal began to muster the financial resources necessary to make investments in key areas of the environment, and to start to address the consequences of years of environmental neglect.

The Evolution of Environmental Issues

Somewhat paradoxically, and in contrast to what might be expected, there is a historic tradition of concern for the environment in Portugal. Ancient laws that persisted into the nineteenth century are good examples of this environmental legacy. Laws that aimed to restrict hunting and the cutting down of certain tree species in specific locations were strongly enforced, with potential punishment including several years imprisonment in the African colonies. At the end of the nineteenth century legislation was also created to protect public access to the coast under the so-called 'Public Maritime Domain'. This far-sighted legislation banned private ownership of buildings and land anywhere within a zone lying up to 50 metres above the maximum high tide mark. In fact such past legislation reveals that, as is the case today, it has often not been the lack of specific legislation but rather its implementation that is the key limitation upon environmental improvement.

It is only comparatively recently that real credibility has been given by the Portuguese State to environmental concerns. In 1971, the National Commission for the Environment (CNA – *Comissão Nacional do Ambiente*), was the first official organisation to be created specifically to deal with environmental issues. The establishment of such a body resulted from the need to prepare for Portugal's participation in the first United Nations Conference for the Environment held in Stockholm in 1972. It is notable that institutional credibility for the environment emerged only in response to external pressures and not as the result of sustained insistence from internal forces. The CNA was responsible for the first genuine attempts at environmental education in Portugal and for producing several important publications on environmental issues. Indeed the organisation was responsible for the first studies that could be considered truly environmental and created an 'Atlas of the Environment' which brought together information from several different areas into one single publication. On the downside, the CNA acted merely as a consultative body without any kind of political power to implement decisive actions (Pimenta and Melo,1993).

In 1974, following the April revolution and the political change that accompanied it, a new framework for dealing with environmental issues emerged through the creation of the Secretary of State for the Environment (SEA – *Secretaria de Estado do Ambiente*) a political body integrated into government. The SEA had greater authority and powers than the CNA, although these powers remained limited in extent. Despite the existence of SEA it was clear that there remained a problem of how to integrate the issue of the 'environment' into general state policy. This was exemplified by indecision surrounding where SEA should best be located and its subsequent movement between several Ministries. During this period the architect Ribeiro Telles emerged as one of the first individuals to draw attention to the environment as an important issue within Portugal. Ribeiro Telles helped to spotlight several major environmental problems and shifted the issue of the environment into mainstream consideration.

In 1985, SEA became the Secretary of State for the Environment and Natural Resources (SEARN – *Secretaria de Estado do Ambiente e Recursos Naturais*), and its closer integration into the powerful Ministry of Planning and Administration of the Territory (MPAT), led to a qualitative change of affairs. Not only was the financial budget for environmental issues multiplied several times over, but SEARN also became responsible for key national policy areas including water resources and protected areas.

In 1987 a further important milestone in the history of the Portuguese environmental movement took place. The approval of the Structural Law for the Environment (LBA – *Lei de Bases do Ambiente*) put in place a basic change with respect to the whole structure of the environment, as well as the principles under which basic environmental laws would be administered and implemented in the future.[1] Unfortunately some of these principles have remained largely rhetorical and still remain to be incorporated into legislation. In the late 1980s, Carlos Pimenta (Secretary of State for the Environment) began to play an increasingly influential role as one of the officials responsible for the LBA, as well as through his decision to move ahead with the demolition of hundreds of illegal houses built along the Portuguese coastline. In response to the increasing profile of environmental issues, in 1990 the environment was split from the Ministry of Planning and given the status of its own Ministry. Whilst it maintained the same powers as before, its newly established status meant a more active role was expected in future political decisions and it was given its own seat in the country's Council of Ministers. Three years later in 1993, the 'National Network of Protected Areas' was created providing for the first time the opportunity to develop a new management policy for all of the Protected Areas designated in Portugal.

Effective environmental policy requires the production and dissemination of information in order to monitor environmental quality and encourage public participation. Although the LBA anticipated these information needs, in reality many of the measures needed to achieve them were never implemented. One example of this was the publication in 1991 of the 'White Book for the Environment', a report on the state of the environment by the Ministry of the Environment with the help of academics and professional experts (Ministério do Ambiente, 1991). The independence of this publication and the exhaustive information compiled within, meant this was a key document intended to provide a baseline study for subsequent regular updating. Although the LBA stated that such a publication should be produced regularly every three to five years, to date only the 1991 edition was ever produced. On an annual basis, a 'Report of the State of the Environment' is published, but methodologically this report lacks balance and is inconsistent in its coverage of key subjects. Furthermore, year to year changes in available data make valid temporal comparisons a near impossibility.

Another lost opportunity relates to the publication in 1994 of the 'National Plan for an Environmental Policy'. This plan for the first time succeeded in involving all of the parties who had a direct interest in the environment, ranging from the services of the Central Government to the local authorities' professional experts, as well as officials from non-governmental organisations (NGOs). The result was a document that not only provided a detailed diagnosis of the state of the Portuguese environment but also outlined some important guidelines for future action. Unfortunately, the submission of this document occurred at the same time as the National Assembly elections of October 1995. In the ensuing political change the document was largely forgotten and a real opportunity to move forward the country's environmental situation remained unrealised.

As Figure 7.1 illustrates, the Second Community Support Framework (CSF II), 1994-99, played a central role in the overall development of the country's environmental policy. The major investments made in basic infrastructures were only possible thanks to European funding (Martins, 1999). More than 720,000 million escudos were invested in environmental projects under the CSF II, with more then 60% of these funds coming from European funds and less than one seventh from central government.

Also critical to the development of the environmental agenda within Portugal has been the increasing role played by non-governmental organisations (NGOs) in promoting awareness of environmental issues. These organisations represent some of the most effective means available for citizens to organise themselves and demonstrate their opinions.

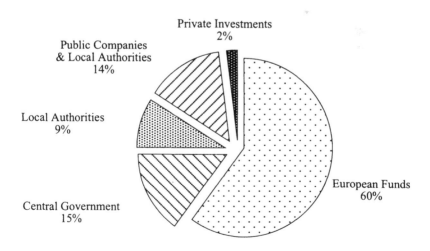

Source: Ministério do Ambiente, 1998

Figure 7.1 Sources of Investment in Environmental Projects under Second Community Support Framework, 1994-99

The importance of environmental NGOs started to grow during the 1980s. From a particular peak in 1987, when 42 new bodies were created, subsequent years have seen the regular creation of new groups (see Figure 7.2). By 1996 there were 150 Environmental NGOs registered in Portugal with a membership of 120,000 individuals. Although initially such groups were regarded as marginal to policy debate, as they have developed their role and their membership has expanded to include academics, researchers and specialists, NGOs have gained in credibility and are increasingly seen as providing an important source of environmental expertise.

The oldest of the environmental NGOs, the *Liga para a Protecção da Natureza* (LPN), was founded as far back as 1948. However it was only in the 1980s that it became more accessible to the general public, previously being mainly a group of university academics with little influence. The other two main NGOs, Quercus and GEOTA, along with the LPN, now play an important role in alerting public opinion, pressurising the government into action and presenting alternative solutions to environmental problems. Partly as a result of these NGO groups, public participation in, and awareness of environmental issues has grown substantially in Portugal in the last 25 years, albeit from a very low base (see Box 7.1). Fundamental to this increased environmental awareness and activity has been higher levels of prosperity among the general population. As basic needs have been met, the desire for an improved quality of life has become central to the aspirations of citizens and hence to wider political debate.

Despite these changes, environmental planning in Portugal continues

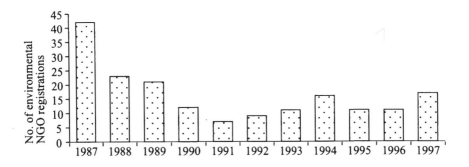

Source: Instituto de Promoção Ambiental, 1998

Figure 7.2 Creation of Environmental NGOs, 1987-97

Box 7.1 Public Participation and Awareness: the Case of Co-incineration

Due to an urgent need to resolve the problem of the accumulation of dangerous waste materials, the Portuguese Government decided to choose two (from four) cement factories to incinerate dangerous wastes under the guidance of special safety measures. However, the whole procedure of choosing the two factory locations was fundamentally flawed, displaying ambiguity and a lack of information. As a result, the process provoked strong public response from those local populations surrounding the cement factories under consideration. Despite statements that several anti- pollution measures would be taken, the local populations, fuelled by distrust of the cement factories given their past negligent behaviour, refused to accept the location of co-incineration near to their homes. Public demonstrations and other lobbying pressures were decisive in forcing the government to appoint an independent scientific commission and suspend the final decision on the incineration sites until after the general election of October 1999. In the resulting report, the independent commission (two scientists appointed by the Portuguese Universities and one by the Ministry of the Environment) concluded that the process should go ahead and identified the location of two cement factories for inciner-ation, one in a new location further from population centres but within a Natural Park (Parque Natural da Arrábida).

This outcome led to further protests from environmental groups and the creation of a second commission to focus on the safety issues related to the health of the surrounding populations. Opposition to co-incineration also resulted in a local boycott of the 2001 Presidential elections in Souselas. These delays in the planning process, electoral boycotts, and the appointment of two independent commissions in response to public pressure, reflects the extent to which the environment has now become a focus for public debate and protest. It would have been difficult to foresee such an outcome even in the mid 1990s, not least because protests have been directed against the cement industry, one of the strongest lobbying groups found in Portugal (GEOTA, 1998).

to exhibit a limited capacity to affect political decisions. Indeed, environmental policy still fails to encourage widespread public participation in the planning process (Soromenho Marques, 1999). The Ministry of the Environment has limited political power with major political and economic lobby groups retaining the ability to influence key environmental decisions in a manner which displays little concern for environmental matters (see Box 7.2). The political situation is further complicated given that the environment forms part of the decision making process across a number of ministries. Whilst it might appear logical that issues such as mineral resources, hunting, and forestry should be under the control of the Ministry of Environment, this is not the case. Given this context, what is necessary is for all ministries to have environmentally sensitive decision-making processes; a situation which for the moment remains something of a distant vision. However, one encouraging step taken by the newly elected government in October 1999 was to place territorial management under the jurisdiction of the Ministry of Environment, a change which been called for by environmental NGOs and experts for many years.[2]

The Main Environmental Problems in Portugal

The 1998 inquiry by the *Observa* Project, a joint study between the Ministry of the Environment and the *Instituto Superior de Ciências do Trabalho e da Empresa* (ISCTE), helped launch a national debate on environmental issues. When members of the public were asked to identify the two major problems that concerned them most, the environment ranked in 5th place. Moreover, 13.5% of the Portuguese identified the environment as one of the two main problems affecting their daily lives. In particular, respondents focused upon pollution, degradation of landscapes, and the lack of proper spatial planning. These results provide further evidence that the Portuguese population has developed a more pronounced environmental conscience.

The 1990s witnessed some major improvements with regard to certain environmental problems relating to access to basic infrastructures such as water availability and the management of waste and sewage. In the early part of the 1990s the situation in Portugal remained very poor, especially when compared with the rest of the European Union (see Figure 7.3). In practical terms only 64% of population had access to water supply and less then a quarter were served by sewage treatment stations. The dramatic increase in the proportion of the Portuguese population with access to the collection of domestic waste, treatment of sewage, and access to water facilities across the 1990s was largely the result of the availability of EU funds. To this end CSF II outlined a series of clear targets which in the case of the environment focused on the provision of basic facilities and trying to attain what were considered to be 'European levels' of access (see Figure 7.4). By the end of CSF II although these targets were not achieved marked improvements were evident. By 1999, 90% of population had a domestic water supply and 55% were served by sewage treatment systems. In addition there were significant improvements in the treatment of industrial wastes, with 62% of this waste disposed of by sanitary burial, incineration or physical/chemical treatment by 1997 (Ministério do Planeamento, 2000).

Water Access and Quality

Access to water and water quality, are two of the country's most serious environmental problems. Although Portugal is a rich country in terms of the overall provision of water, the main problem lies in terms of its distribution over time and space. While the north of Portugal is more then self-sufficient in water, the south of the country often has serious difficulties in meeting the water needs of local populations, especially in the summer months. Moreover, the use of water resources does not always proceed in a rational

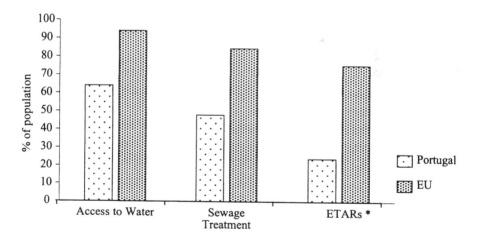

Source: Direcção Geral de Desenvolvimento Regional, 1994

* Sewage Treatment Stations

Figure 7.3 Basic Infrastructures in Portugal and the European Union, 1991

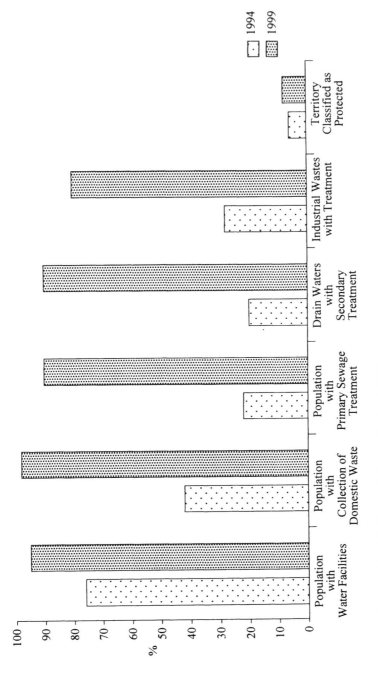

Source: Direcção Geral de Desenvolvimento Regional, 1994

Figure 7.4 Environment Targets under the Second Community Support Framework 1994-99

manner. For example it is estimated that the agricultural industry wastes enough water in just one year to supply the population of the two major cities of Lisbon and Porto.

Water quality is also problematic. An analysis of Portuguese river basins undertaken between 1996-97 demonstrated the profound levels of existing pollution (see Figure 7.5). More then 38% of rivers were classified with high levels of pollution meaning that no use (besides minor navigation and some irrigation) could be made of them. Only 20% were classified with low levels of pollution, which made them suitable for all commercial and industrial uses. Levels of water deterioration varied as a result of industrial activity and agro-food practices such as cattle rearing, olive oil factories and the pig industry.

Water related problems are once again concentrated near the coast, close to the industrial growth poles and major urban centres. Although the situation is beginning to improve, there remains a large amount of clean-up work to be undertaken. To maintain progress requires a continued change in the attitudes of Portuguese industry towards the environment. This can only happen with more vigilance and a stricter observation and interpretation of the law. In line with the 'polluter pays principle', a principle established by the LBA and operational since 1995, industries that polluter should be made to pay for the clean up operation. However, in order to give this measure real

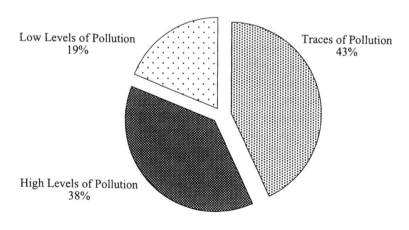

Source: INAG, 1998

Figure 7.5 Portuguese Water Quality, 1996-97

force, the price paid must be higher than the profits obtained from non-observance of environmental regulations; something which is often not the case currently.

Waste Management

One of the most critical environmental issues within Portugal concerns the management of domestic waste. Up until the mid-1990s there were more than 300 dumpsites in Portugal which were responsible for disposing of nearly 30% of solid domestic waste (see Figure 7.6). All of these were operating under very poor conditions. Most lacked any kind of impermeable layer to prevent leaching and this led to widespread contamination of soils and underground water resources. In response a programme was developed to close all 300 sites by the end of 1999 and replace them with proper landfill schemes. This achievement marked a major improvement in the quality of life of the local populations as well as for those who depend on underground water resources; that is the majority of the Portuguese population (Direcção Geral do Ambiente, 1996).

Major investments have also been made in the system of sewage treatment. Whilst it is true that hundreds of sewage treatment stations (ETAR's) have been built, there is evidence to suggest that a significant number are either not in operation, or are not working properly. Several reasons account for this situation; partly due to construction problems partly to the non-existence of financial support to help maintain these structures. The lack of monitoring and regulation of these treatment stations is also problematic given that it is cheaper for them to be switched off for the majority of the time. As a result there remain problems with river pollution whilst a significant proportion of sewage is still being released into the ocean without having undergone any significant treatment process. (Direcção Geral do Ambiente, 1998).

Rural Land Use

As soils for agricultural purposes are generally poor within Portugal, a significant proportion do not possess the physical conditions to produce good quality crops. As a result, many rural areas suffer from high levels of erosion, an excess usage of fertilisers and an increasing process of desertification. Afforestation is also important in this respect. The large scale planting of certain tree species, notably eucalyptus, on inappropriate soils is a major cause of soil degradation nationally. Following the introduction of these species it is common to observe a dramatic change in the landscape whilst, even more importantly, water availability within the soil is reduced.

Poor management of forested areas combined with rural depopulation has also led to a recent increase in the number of forest fires. As Figure 7.7 illustrates, forest fires are responsible for the loss of hundreds of hectares of land each year in Portugal despite the major investments made annually in both human and material resources, to help fight them. Given that forestry is an important component of economic development, investment is needed in preventative measures, rather than simply in fire-fighting systems and equipments. Relevant authorities need to become much more proactive in promoting prevention rather than simply being reactive to the fires themselves. However, the situation is particularly difficult to deal with due to the fragmentation of forested property. Scattered ownership means that attempts to prune and plant trees properly, measures which are crucial for reducing the number of forest fires, are not being implemented effectively (Ministério do Ambiente, 1991).

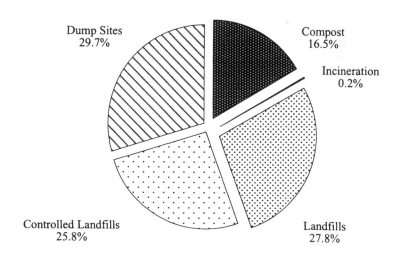

Source: Ministério do Ambiente, 1998

Figure 7.6 Disposal of Domestic Solid Waste in Portugal, 1996

Planning and Spatial Management

Arguably the most important environmental problem in Portugal at the present moment is the lack of any co-ordinated spatial management of environmental issues by government (see Box 7.2). This, in turn, is a problem rooted within the ever-increasing occupation of the coastal zone by the Portuguese population. The complex planning problems of the two main metropolitan areas of Lisbon and Oporto further aggravates difficulties. Pressure on municipalities to boost their income from taxes and licenses from new building developments has led to current Municipal Plans (*Planos Directores Municipais*) containing enough planned new housing for around 20 million people in the period to 2010 (Schmidt, 1999). Given a current population of 10 million and little projected population growth, such planned overcapacity reflects the development of a culture of land speculation driven by the powerful building and construction lobby.

Poor spatial management is also evident with respect to the growing

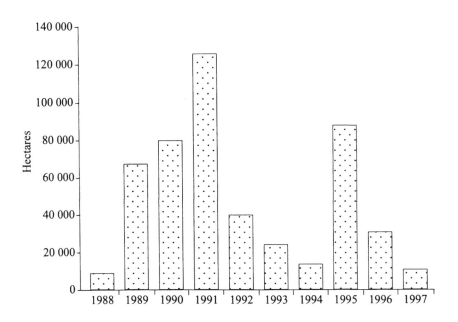

Source: Direcção Geral das Florestas, 1998

Figure 7.7 Areas Burnt by Forest Fires in Portugal, 1988-97

Box 7.2 The Vasco da Gama Bridge

Many of the environmental problems in Portugal either result from, or are exacerbated by, the decisions of Central Government. A good example is the construction of the Vasco da Gama Bridge across the Tagus in Lisbon, which reveals the lack of power and political influence of the Ministry of the Environment vis-à-vis other government Ministries. Although one of the main goals of the construction of this new bridge was to divert traffic from the existing '25th of April Bridge', the chosen location did not satisfy this objective. In fact many of the decisions taken in relation to the construction of the 'second bridge' across the Tagus proved to be highly controversial from an environmental standpoint. First, there was no prior Environmental Impact Assessment (EIA) undertaken for any of the three alternative locations; only after one had been chosen was an EIA actually carried out. Second, the chosen route was the one designed to divert the least amount of traffic from the existing bridge. Third, the preferred route was the one nearest to the Tagus Estuary Natural Reserve, widely considered to be one of the most important bird sanctuaries in Europe. Finally, the chosen route has been responsible for bringing new urban development pressures to an area which had previously been free of them, creating new suburbs in the east of Lisbon whilst failing to solve existing problems.

The above sequence of decisions occurred in spite of the opinions of the Ministry of Environment. The Ministry's submission to the planning inquiry stated that:

> ... the Environment Impact Study made for the Ponte Vasco da Gama does not justify the project. The goals stated as the basis for the construction of a new bridge will not be achieved and negative impacts on both the environment and the management of the territory will be considerable, causing significant negative effects on the Tagus estuary, terrestrial ecosystems and environmentally sensitive areas (Ministério do Ambiente Recursos Naturais, 1994, p.35).

However these views, as well as those of various environmental NGOs, were ignored by the powerful Ministry of Public Works, Transport, and Communications, which decided to go ahead and build the bridge in their preferred location. As predicted, the Vasco da Gama bridge, which opened in 1998, did not ease congestion on the '25th of April Bridge' and a third bridge across the Tagus is now under active consideration. (Joanaz de Melo,1998).

human occupation of the coastal fringe. This trend has led to a situation where the sea now threatens several poorly located human settlements. In a number of places the solution has been to hard engineer solutions that, in many cases, have not solved the problems but merely displaced them to elsewhere on the coast. Improved coastal management in these circumstances is important for a number of reasons. First, these kinds of solutions are extremely expensive. Between 1996-99, the budget for hard engineering coastal works was around 6,000 million contos. Second, solutions have often been designed to protect private interests in areas where they should never have been permitted to locate in the first place. Finally, much of this intervention has happened without appropriate prior studies; a lack of preparation that is now reflected in the unsatisfactory results of many of these coastal defences (Ministério do Ambiente, 1998).

More positively, a number of changes are starting to emerge with regard to coastal zone management. Studies of the evolution of the coast of Portugal are now in progress and these should provide the basis for the implementation of 'Integrated Coastal Zone Management' techniques in the future. Official bodies have begun to move away from large investments in hard coastal engineering works and implement new plans for the management of coastal areas (POOCs – *Planos de Ordenamento da Orla Costeira)*. The zone effectively covered by these plans is a fringe of 500 metres above the high tide mark where many environmental problems are concentrated. However, to date the results of the first POOCs has been less than satisfactory. In practice the plans have sought to avoid conflicts with municipalities and several of them are little more than localised beach plans, designed to control the concessions and activities found there. Once again, it is not the lack of legislation that is the problem rather the failure of effective implementation.

Environmentally Protected Areas

Since 1993 the designation of Protected Areas has been organised into a national structure comprising; one National Park, eleven Natural Parks, eight Natural Reserves, three Protected Landscapes, ten Classified Sites, and five National Monuments (see Figure 7.8). These protected areas together occupy more then 651,000 ha. (around 7% of the total territory) with a population of 200,000 inhabitants. Despite the comprehensive nature of this classificatory system for protecting natural areas, policy to manage these areas effectively remains weak. To date, policy has focused mainly on day to day management issues and short term change, with little attention to longer

Box 7.3 Protected Areas: the Natural Park of SW Alentejano e Costa Vicentina

The Natural Park of the *SW Alentejano e Costa Vicentina,* south of Sines, is one of the best preserved stretches of coastline in Portugal and exemplifies many of the problems associated with protected areas in Portugal. Originally classified in 1989 as a protected landscape, its status was changed to that of a Natural Park in 1995 in order to strengthen the degree of protection. Traditionally the area has been relatively free from the pressures of tourist development and the natural landscape was well preserved. However in the 1990s developers put forward plans for more than 50,000 tourist beds in an area of less than 50 kilometres of coastline; plans which could have potentially devastated the region's natural environment (Schmidt, 1999). Fortunately these plans were never implemented. Only a small number of these beds were built and almost all were scattered second homes rather than the small-scale tourist resorts originally proposed. At the same time, a mega-agricultural project was developed at Herdade do Brejao, located in the middle of the park, under the pretence that it was an organic farming project. Against the opinions of the park authorities development went ahead with the support of central government. When the project eventually failed a few years later it left a legacy of polluted soils and the wreckage of old and derelict greenhouses. Even though it was not responsible for the situation, the authorities of the Natural Park were left to rectify the situation, a process that took several years to achieve.

Somewhat ironically this area now suffers from a surfeit of plans, with new regional, local and special plans all recently created. The Alentejo Coast, which includes the Natural Park of the SW Alentejano and Costa Vicentina, now operates under the jurisdiction of at least four different plans: the Regional Plan for the Alentejo Coast (PROTALI); Municipal Lead Plans (PDMs); the Management Plan for the Coastal Fringe (POOC); and the Management Plan for the Natural Park (POPNSWCA). Rather than solving problems this situation is creating new ones, with clashes of interest and ambiguities arising out of the lack of articulation between different plans. Although the coast has been preserved to some extent, there is a clear need for improved co-ordination and policy integration if actions are to be more effective in the future.

term strategic objectives. Partly this is the result of a lack of finance, with small budgets fundamentally restricting the scope of management activity. However, much of the planning that has taken place within these areas has focused upon imposing restrictions without providing adequate alternatives to local populations. For this reason it is perhaps not surprising that many residents within Protected Areas have developed a degree of animosity towards them.

In response to these types of problems, legislation in 1996 introduced incentives for investment within Protected Areas aimed at creating more employment and maintaining population levels especially in interior regions. There is increasing recognition that these areas cannot be viewed simply as islands of conservation within the national territory, but rather must be integrated into their surrounding regions so that they are viable places to live and work (Ministério do Ambiente e Recursos Naturais, 1998). One key element in this respect are current trends towards growing tourist demand in rural areas (see Williams, Chapter Four). Protected Areas are increasingly attractive destinations for tourists but as a result are beginning to suffer the consequences of higher levels of visitors which, in the absence of strong management, is liable to lead to further environmental problems (see Box 7.3).

Conclusions

The analysis presented here portrays a certain paradox with regard to environmental issues within Portugal. Although it is true that serious environmental problems exist and pressures on the environment are growing, it is also true that a number of significant improvements have been made in recent years. Overall, the quality of life for the vast majority of the Portuguese population has improved with respect to access to basic services such as water, sewage treatment, and rubbish collection. There has been major progress in the treatment of domestic waste, thanks mainly to the National Plan that led to the removal of one of Portugal's most serious environmental problems, that of unregulated dumpsites. Integrated coastal zone management plans, along with the Regional Plans for coastal management, provide a useful starting point for improvement, as do the specific studies of the coast and the production of risk maps. The existence of a nature conservation policy places greater value on protected areas and no longer treats them as isolated islands within overall national territory. Despite a wait of ten years and much controversy, the publication of the 'Structural Law for

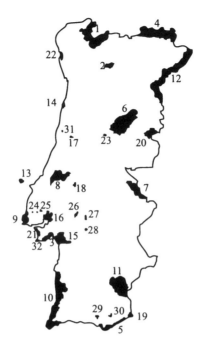

1 P.Nacio. Peneda Gerês
2 P.Nat. do Alvão
3 P.Nat. da Arrábida
4 P.Nat. de Montezinho
5 P.Nat. da Ria Formosa
6 P.Nat. Serra da Estrela
7 P.Nat. Serra S.Mamede
8 P.Nat. Serras de Aires e Candeeiros
9 P.Nat. de Sintra Cascais
10 P.Nat. do SW Alentejano e Costa
   Vicentina
11 P.Nat. do Guadiana
12 P.Nat. do Douro Internacion
13 R.Nat. da Berlenga
14 R.Nat. Dunas de S. Jacinto
15 R.Nat.do Estuário do Sado
16 R.Nat. do Estuário do Tejo
17 R.Nat. do Paul da Arzila
18 R.Nat. do Paul do Boquilobo

19 R.Nat. do Sapal de Castro Marim e
   V.R.S.António
20 R.Nat. da Serra da Malcata
21 Paisagem Prot. da Arriba Fossil da Costa da
   Caparica
22 Paisagem Prot. do Litoral de Esposende
23 Paisagem Prot. da Serra do Acor
24 Sítio Classificado Campo de Lápias da
   Granja dos Serroes
25 Sítio Classificado Campo de Lápias de Negrais
26 Sítio Classificado do Centro Historico de
   Coruche
27 Sítio Classificado da Agolada
28 Sítio Classificado do Acude do Monte da Barca
29 Sítio Classificado da Rocha da Pena
30 Sítio Classificado da Fonte da Benemola
31 Sítio Classificado Montes de Sta. Eulalia e
   Ferestelo
32 Sítio Classificado da Gruta do Zambujal.

Abbreviations

P. Nacio	National Park		
P. Nat	Natural Park		
R. Nat	Natural Reserve	Paisagem Prot.	Protected Landscape
		Sítio Classificado	Classified Site

Source: Instituto de Conservação da Natureza (ICN), 1999

Figure 7.8 Network of National Protected Areas, 1999

the Management of the Territory' in 1998 provided important new foundations for the development of environmental policy, although it fell far short of what could have been achieved. The relocation of responsibility for territorial management to the Ministry of the Environment is also a further positive step.

Despite these achievements, the scale of current environmental problems remains substantial. The National Plan for Social and Economic Development, which provides an outline of the country's economic and social policy for the 2000-2006 period, identifies a series of key objectives with regard to the environment:

- improvements in water quality.
- reinforcement of the 'three R's' policy for dealing with industrial and urban solid waste (i.e. 'Reduction, Reutilization and Recycling').
- evaluation of the coastline through the implementation of specific plans and the development of basic studies to improve the knowledge of natural systems.
- implementation of measures to improve the quality of life in urban areas.
- intervention in highly sensitive areas, such as those which due to improper agricultural use are suffering from desertification.
- implementation of a nature conservation scheme across 2,000 sites and the improvement of the quality of life of the populations that live inside protected areas in order to counteract depopulation.
- integration of environmental concerns in decisions taken by other areas (e.g. industry, agriculture, energy, transport, and tourism).

Given these laudable objectives, the greatest challenge is how to develop the capacity for implementing an aggressive and effective environmental policy within Portugal. Protection of the environment faces the problems of conservation within a context of rapid and non-sustainable economic growth that pays scant attention to environmental costs. Interventions from several bodies within the public administration (e.g. public works, tourism, forestry, industry and economy) are responsible for many of the negative impacts upon the environment, even inside protected areas. Clearly, it is crucial to try to change this negative role of governmental and public bodies and move towards some form of integrated and sustainable environmental policy. Although existing environmental policy in Portugal is principally concerned with conservation, designed ostensibly to protect, it must in the future become more directed towards policies of sustainable development. Within such a policy it is necessary to sustain certain traditional activities, such as agriculture, as these are largely responsible for the 'natural landscapes' which currently exist. However these landscapes are also an important economic resource, especially in a country where tourism is a major economic activity, and consequently must be managed accordingly.

In response to the serious environmental problems discussed within this chapter, environmental education and awareness in Portugal is developing. The impact of this will take time to feed through into the political agenda, but a population more aware and concerned about the environment does create new possibilities for tackling the negative heritage arising from past actions and avoiding these problems in the future. Current and future investments being made in the environment through large amounts of EU funding provide a unique opportunity to pursue an environmental agenda. Given that such financial resources are unlikely to be available again in the future it is imperative that effective use is made of these resources in the early years of the twenty-first century.

Acknowledgements

The author is grateful for all the help given by Dr. Martin Eaton in the preparation of this article.

Notes

1   The *Lei de Bases do Ambiente* (LBA - Structural Law for the Environment) is referred to within the Portuguese Constitution and defines the basic environmental principles from which all subsequent environmental legislation is developed. For example the LBA states that there should be a national network of protected areas and this resulted in legislation, DL 19/93, which provides the specific regulations concerning the creation and management of a network of protected areas in Portugal.

2   The incoming government in 1999 created a new ministry, *Ministério do Ambiente e Ordenamento do Território* (MAOT – Ministry of Environment and Territorial Management), which brought issues of territorial management formerly within the *Ministério do Equipamento, do Planeamento e da Administração do Território* (MEPAT - Ministry of Equipment, Planning and Territorial Management) into the Ministry of the Environment.

References

Direcção Geral do Ambiente (1996), *Relatório de Estado do Ambiente, 1995*, DGA, Lisbon.

Direcção Geral do Ambiente (1998), *Relatório de Estado do Ambiente, 1998*, DGA, Lisbon.

Direcção Geral do Desenvolvimento Regional (1994) *Community Support Framework for Portugal: Annual report 1993*, DGDR, Lisbon.

GEOTA (1998), 'Despacho de Co-incineração' at www.despodata.pt/GEOTA\Co incineração \despache.htm

Joanaz de Melo, J. (1998), 'The New Bridge Over the Tagus Estuary: How Not to Develop a Project', at www.despodata.pt/geota

Martins, V. (1999), 'O Estado do Ambiente em Portugal: Balanços e Perspectivas', in Conselho Económico e Social (ed) *Ambiente e Economia e Sociedade*, Serie Estudos e Documentos, Conselho Económico e Social, Lisbon, pp.29-82.

Ministério do Ambiente (1991), *Livro Branco sobre o Estado do Ambiente em Portugal*, MARN, Lisbon.

Ministério do Ambiente (1998), *Balanço da Acção Governativa do Ministério do Ambiente*, MA, Lisbon.

Ministério do Ambiente e Recursos Naturais (1994), *Parecer da Comissão de Avaliação do MARN sobre o EIA da Nova Travessia Rodoviaria do Tejo em Lisboa*, MARN, Lisbon.

Ministério do Planeamento (2000), *Quadro Comunitário de Apoio III: Portugal 2000-2006*, Ministério do Planeamento, Lisbon.

Pimenta, C. and Joanaz de Melo, J. (1993), *O Que é a Ecologia*, Difusão Cultural, Lisbon.

Schmidt, T. (1999), *Portugal Ambiental: Casos e Causas*, Celta Editora, Oeiras.

Soromenho Marques, V. (1999), 'O Estado do Ambiente em Portugal: Uma Perspectiva Crítica sobre a Política de Ambiente em Portugal', in Conselho Económico e Social (ed) *Ambiente e Economia e Sociedade*, Serie Estudos e Documentos, Conselho Económico e Social, Lisbon, pp.83-96.

8 National Political Change in Portugal, 1976-99

RICHARD A.H. ROBINSON

Introduction

Writing on the eve of the 25th anniversary of the military coup of 25 April 1974, which ended 48 years of authoritarian rule in Portugal, but not writing with Portugal in mind, the columnist Joe Rogaly (1999) described the current state of democratic politics in the West thus:

> We are in the age of micropolitics. Party labels no longer stand for clear ideologies or sets of values. They have become logos, means of identifying alternative troupes of office-seekers. We do not elect visionaries, but actor-managers: not great thinkers who will try to change the world but good TV performers whom we can rely on to tinker with details.

If Rogaly's description of 'micropolitics' were to be transferred to the current state of politics in Portugal, it would remain essentially accurate but the distance traversed in the last quarter-century to get to this stage would be even more remarkable. Whereas the well-established Anglo-Saxon democracies have evolved from the ideological conflict of the Cold War and debates about different ways forward from the recession provoked by the oil price rise of 1973, Portugal has evolved into a functioning constitutional democracy from the colourful military-dominated confusion of the PREC (*processo revolucionário em curso*), when the hegemonic discourse was socialist and Marxist and the chances of democratic consensus sometimes seemed slim. Outsiders wondered whether a stable democracy could ever be achieved as the four coups, or attempted coups, and the six provisional governments of April 1974-July 1976 gave way to the five unstable constitutional governments of July 1976-October 1979, which led to elections before the first four-year constitutional term for the legislature had run its course.[1]

In this chapter the transition from the seemingly ideologically-dominated politics of the mid-1970s to the voter-oriented political marketing of the late 1990s is examined with respect to the following dimensions: changes in the constitutional framework of Portuguese politics; electoral evolution at the national level; the evolution of the major political parties; and continuing preoccupations with the health of Portuguese constitutional democracy.

The Evolution of the Constitutional System

The Constitution of the Portuguese Republic, running to 312 articles, was passed on 2 April 1976 by the Constituent Assembly elected on 25 April 1975. Only the Deputies of the Christian-Democrat CDS *(Centro Democrático Social* – Democratic Social Centre) voted against. To get to this stage the political parties that emerged in the wake of the coup of 25 April 1974 had had to agree to commitments with the left-dominated Armed Forces' Movement (*Movimento das Forças Armadas* – MFA) to be able to take part in the elections for the Constituent Assembly and to pass the final draft of the Constitution. At least up to the turning-point of the military confrontation of 25 November 1975, in which the extreme left was defeated by the pro-democratic forces, the Constitution's key articles had been drafted and debated against a background of a revolutionary process powered by the Communists of the pro-Moscow PCP (*Partido Comunista Português* – Portuguese Communist Party) and the small groups to their left in alliance with the MFA, whose key more moderate leaders broke with their pro-Communist and more revolutionary colleagues in August 1975 (Manuel, 1995; Robinson, 1979).

The Communists and other revolutionaries promoted slogans such as *O Povo está com o MFA* ('The people is with the MFA') and the Chilean-derived *O Povo, unido, jamais será vencido* ('The people, united, will never be defeated'), and songs like *Soldado, amigo, o Povo está contigo,* ('Soldier, friend, the people are with you'). The results of the elections of April 1975 (see Table 8.1) demonstrated that this interpretation of 'the People' represented only a fifth of the registered electorate. However it should be remembered that the Socialist Party (PS – *Partido Socialista*) of Mário Soares, the main beneficiary of these elections, although in increasingly bitter competition with the Communists from late 1974, also officially proclaimed itself a Marxist party.[2] Against this background, and given that the Deputies of the PS and PCP constituted over half the Assembly, it was not altogether surprising that the Constitution of 1976 reflected ideological ambiguity. This ambiguity was epitomised in the text of its Article 2:

> The Portuguese Republic is a democratic State, based on popular sovereignty, on the respect for and the guarantee of fundamental rights and liberties, and on the pluralism of democratic expression and political organisation, which has as its objective assuring the transition to socialism by means of the creation of conditions for the democratic exercise of power by the working classes.

The reference to socialism and the status quo after the revolutionary implementation of land redistribution in the Alentejo and the nationalisation of all but foreign-owned banks and insurance companies and their shareholdings in 1975, was reinforced by specific commitments to the irreversibility of these revolutionary gains (Article 83) and the development of the revolutionary process to achieve further collective appropriation of the principal means of production (Articles 10, 80 and 90).

Alongside the rhetoric of socialism, the Constitution set out the essential balance of powers between what were termed the five 'organs of sovereignty': the President of the Republic, the Council of the Revolution, the Assembly of the Republic, the Government, and the Courts. The Council of the Revolution, chaired by the President of the Republic and consisting of officers from the three armed services, was a product of the Platform of Constitutional Agreement of the 26 February 1976 between the MFA and the major political parties. Its essential functions were to act as a watchdog to ensure that the provisions of the constitutional text were strictly adhered to, to give general political advice to the President, and to be the controlling body for the armed forces, guaranteeing 'fidelity to the spirit of the Portuguese Revolution of 25 April 1974' (Articles 142-149). Thus it was the continuation of the MFA by constitutional means. Outside the military sphere the independence of the courts, subject only to the law, was guaranteed (Article 208).

In the circumstances of the time it was thought politic to check the powers of the President of the Republic, the government and the legislature in an effort to prevent the dangers of charismatic leadership and the unpredictability of parliaments elected by proportional representation. The result was that the President of the Republic was elected by universal suffrage of those over 18 (the President had been popularly elected in undemocratic

Table 8.1 Results of Elections to the Assembly of the Republic, 1975-99

Poll Date	UDP and allies	PCP and allies	MDP	PS	PRD	PPD/PSD	AD	CDS	PSN	ADIM	Total No of seats
24-4-1975											
% of votes	0.8%	12.5%	4.1%	26.4%	–	26.4%		7.6%	–	0.03%	250
N° of seats	1	30	5	81		81		16		1	
25-4-1976											
% of votes	1.7%	14.4%	–	34.9%	–	24.4%		16.0%	–	–	263
N° of seats	1	40		107		73		42			
2-12-1979											
% of votes	2.1%	APU 18.8%	–	27.3%	–	–	AD 45.3%	–	–	–	250
N° of seats	1	47		74			128				
5-10-1980											
% of votes	1.4%	APU 16.8%	–	FRS 27.8%	–	–	AD 47.6%	–	–	–	250
N° of seats	1	41		74			134				
25-4-1983											
% of votes	UDP+PSR 1.2%	APU 18.1%	–	36.1%	–	27.2%		12.6%	–	–	250
N° of seats	–	44		101		75		30			
6-10-1985											
% of votes	UDP 1.3%	APU 15.5%	–	20.8%	17.9%	29.9%		10.0%	–	–	250
N° of seats	–	38		57	45	88		10			
19-7-1987											
% of votes	UDP 0.9%	CDU 12.1%	0.6%	22.2%	4.9%	50.2%		4.4%	–	–	250
N° of seats	–	31	–	60	7	148		4			
6-10-1991											
% of votes	PCP+PEV+UDP+PC(R) 8.9%			29.1%	0.6%	50.6%		4.4%	1.7%	–	230
N° of seats	17			72	–	135		5	1		
01-10-1995											
% of votes	UDP 0.6%	CDU 8.6%	–	43.7%	–	34.1%		CDS-PP 9.1%	0.2%	–	230
N° of seats	–	15		112		88		15	–		
10-10-1999											
% of votes	BE 2.4%	CDU 9.0%	–	44.0%	–	32.3%		8.3%	0.2%	–	230
N° of seats	2	17		115		81		15	–		

Source: STAPE, (various years)

environments in 1918 and from 1928 to 1958), as was for the first time the single-chamber parliament, the Assembly of the Republic.

The President's term of office was five years and incumbents were limited to two consecutive terms. If one candidate for President got more than 50 per cent of the poll, then this candidate would be elected on the first round, as were General Eanes in 1976 and 1980, Mário Soares in 1991 and Jorge Sampaio in 1996. If no candidate achieved 50 per cent in the first round, then the two candidates with the most votes would face each other in a second round, as in 1986, when Soares defeated Freitas do Amaral, leader in the first round, by a slender margin. The Prime Minister and his government were appointed by the President but were responsible to the Assembly of the Republic.

The Assembly of the Republic was to be elected by a closed list system in areas coincident with the administrative districts of the country by the d'Hondt method of proportional representation. The number of Deputies elected by a district is in proportion to its percentage of the national registered electorate and the Deputies are selected in the order they appear on the party list on the basis of the valid votes cast for the particular party lists. Thus, supposing a district was to be represented by seven Deputies, and four party lists (A,B,C and D) of seven candidates stood for election, the selection of successful candidates would be as in Table 8.2. As this Table shows, the votes cast are divided by 2, 3, 4 etc. and the highest seven numbers (underlined) elect a Deputy each. Thus candidates numbers 1, 2 and 3 are elected from List A and the first two candidates on List B and the first two on List C.

This system was chosen for its transparency and fairness and in the expectation that rarely would a single party gain an absolute majority in the

Key to Table 8.1

UDP	União Democrática Popular	PSR	Partido Socialista Revolucionário
PC(R)	Partido Comunista (Reconstituido)	BE	Bloco de Esquerda
PCP	Partido Comunista Português	MDP	Movimento Democrático Português
APU	Aliança 'Povo Unido'	CDU	Coligação Democrática Unitária
PEV	Partido Ecologista 'Os Verdes'	PS	Partido Socialista
FRS	Frente Republicano e Socialista	PRD	Partido Renovador Democrático
PPD	Partido Popular Democrático	PSD	Partido Social Democrata
AD	Aliança Democrática	CDS	Partido do Centro Democrático
PSN	Partido de Solidariedade Nacional		Social
ADIM	Associação Democrática	CDS/PP	CDS-Partido Popular
	Independente de Macau		

Assembly, an expectation which remained true until the PSD (*Partido Social Democrata* – Social Democrat Party) won a landslide victory in 1987 (Table 8.1). Until then the system had put the premium on coalitions: although obtaining less than half the votes cast, the centre-right coalition, Democratic Alliance for a New Majority (AD – *Aliança Democrática*), overcame the PS in 1979 and 1980 and gained an absolute majority of seats. Where no party or electoral coalition has won a majority of seats, the outcome has either been a coalition government, as in the 'centre bloc' of the PS and PSD in 1983-85, or uneasy minority governments in the cases of the PS in 1976-77 and the PSD in 1985-87. Although the PS-based government elected in 1995 did not have the majority of seats in the Assembly, it ran its full term because opinion polls indicated its continuing popularity compared with that of opposition parties, which therefore avoided bringing it down for fear of losing seats in the election that followed.

The legislative elections of 10 October 1999 led to an unprecedented situation.[3] The PS of Guterres won most votes but only half the seats in the Assembly of the Republic, with the combined strength of the other parties also coming to 115 seats, making it impossible (if all PS Deputies remain loyal) for the PS government to lose a vote of censure, but also making it necessary for the government to negotiate support or abstention by other parliamentary groups to get legislation passed. Commentators noted that it was the votes of only 43,040 non-resident electors, the results of which were published last as is the norm, which had produced this curious tie between government and opposition. The constitutional and electoral arrangements from 1976 permit Portuguese living abroad to participate in Assembly elections, with four Deputies representing those who register with Portuguese consulates: two Deputies are elected for Europe and two for the rest of the world. The election of Deputies from Macau and Mozambique, present in the Constituent Assembly, has been discontinued.

Table 8.2　Election of Candidates in a District by the d'Hondt System of Proportional Representation

	List A	List B	List C	List D
Votes Cast	*50 000*	*35 000*	*32 000*	8 000
Divided by 2	*25 000*	*17 500*	*16 000*	4 000
Divided by 3	*16 667*	11 667	10 667	2 667
Divided by 4	12 500	8 750	8 000	2 000

The Constitution of 1976 was innovatory inasmuch as it recognised a special status of regional politico-administrative autonomy for the Atlantic archipelagos of the Azores and Madeira. Although each archipelago has its own regional government and prime minister responsible to a locally elected regional assembly, the degree of autonomy given to the archipelagos is limited in the statutes which have to be approved by the Assembly of the Republic. The operation of regional autonomy is overseen in Ponta Delgada and Funchal by centrally appointed Ministers of the Republic, representing the State. The system has worked quite well and to the advantage of the archipelagos and has headed off the separatist tendencies that surfaced in 1975 during the revolutionary process.[4] Bickering and minor frictions have been constants of the island-mainland relationship as regional prime ministers, particularly the exuberant populist Alberto João Jardim in Madeira, have agitated for central financial support to offset 'the costs of insularity'. After over thirty years in office, Jardim still both irritates and entertains mainland opinion with his outspoken comments and demands for an end to the office of Minister of the Republic and for 'home rule'. In the Azores, where the PS broke the PSD's monopoly on regional government in 1996, regional identity was asserted in the 1980s in disputes over where and when the Azorean flag should be flown in relation to the national flag. Despite occasional frustration in the archipelagos with the limits of autonomy, and occasional signs of exasperation with island behaviour on the mainland, the introduction of politico-administrative autonomy has generally proved positive.

The evolution of the constitutional system, reflecting (with some reservations) political and social evolution in Portugal, can be traced through the four revisions of the text of the Constitution of 1976. The first revision required the consent of two-thirds of the Deputies present, provided these totalled over half the Assembly of the Republic (Article 286 of the text of 1976). The essential change brought about in the first revision in 1982, during the Presidency of Ramalho Eanes and initiated by the centre-right AD coalition under Pinto Balsemão of the PSD, was to remove the military from power by abolishing the Council of the Revolution. The armed forces came under civilian control through the subsequent Law of National Defence, with a Supreme Council of National Defence chaired by the President, who since 1986 has been a civilian.[5] The role of constitutional watchdog passed to the new Constitutional Court consisting of 13 judges, while minor adjustments of the presidential powers within the existing semi-presidentialist framework reflected irritation by the government and the PS leader Soares with Eanes's past conduct rather than major considerations of principle.

The AD coalition had hoped that the first revision would remove the socialist rhetoric and commitment of various articles, but Soares's PS declined to provide the support needed for such a purging. Rightly anticipating that this course of inaction would exacerbate frustrations and rivalries within the AD coalition, the PS took advantage to top the 1983 poll following the messy break-up of the AD coalition. Thus, though sounding increasingly hollow as Portugal negotiated its entry into the European Economic Community (EEC), Article 83 continued to proclaim the irreversibility of the nationalisations of 1974-75. Article 2 was rephrased to omit the working classes but retained its socialist commitment. The Republic was now defined as:

> A democratic law-bound State [*Estado de direito*], based on popular sovereignty, on respect for and the guarantee of fundamental rights and liberties and on the pluralism of democratic expression and political organisation, which has as its objective assuring the transition to socialism by means of the realisation of economic, social and cultural democracy and the deepening of participatory democracy.

Another major revision of the text was agreed in 1989 between the then dominant PSD of Cavaco Silva, eager to slim down the state sector, and the pragmatic PS leader Vítor Constâncio, though this was to be a factor in his resignation after criticism from Soares's supporters. In essence, this second revision of the Constitution removed the veto on reprivatisation and so effectively changed the economic model while keeping elements of social democracy. The Economic and Social Council was introduced as a corporate body where representatives of government, employers and unions could debate and agree objectives. Article 2 was again rephrased, so that the Republic was redefined as:

> A democratic law-bound State [*Estado de direito*], based on popular sovereignty, on the pluralism of democratic expression and political organisation and on the respect for and the guarantee of effectuation of fundamental rights and liberties, which has as its objective the realisation of economic, social and cultural democracy and the deepening of participatory democracy.

Thus the commitment to socialism was at last removed, leaving an ideologically open-ended text more in conformity with the constitutions of European Union (EU) partners. A third but minor revision in 1992 brought the Portuguese text into conformity with the Maastricht Treaty of European

Union, permitting residents from other EU countries to vote in local elections.

The idea of the referendum had made a passing appearance in the text of 1989 (Article 170) but the fourth revision of 1997, agreed by the PS minority government and the opposition PSD, reflected the growing concern among the political elite with the apparent failure to deepen popular participation and combat apathy and cynicism. Although Presidents of the Republic have generally enjoyed consistently high satisfaction ratings in opinion polls, and the popularity of prime ministers has fluctuated in the last twenty years, the approval ratings for governments have been lower. Ratings for the performance of the Assembly of the Republic have been lower still, indeed sometimes so low as to indicate popular contempt for parliamentarians and their works.

Apart from unflattering opinion surveys, an indicator of some loss of democratic good health could be seen in declining voter turn out for legislative elections. As table 8.3 illustrates, there has been a notable decline in turnout to the Constituent Assembly elections from an impressive 91.7 per cent in 1975 to only 61.1 per cent in 1999. Against this apparently established trend there has to be set the failure of the electoral register to keep up with deaths and migration. The register was finally revised for the October 1999 election revealing a total register of 8,845,179 names compared with 9,114,338 in June 1997. However despite this, only 61.1 per cent of the revised electorate turned out on 10 October 1999, after a lacklustre campaign overshadowed by the violence in East Timor following the referendum of 30 August and, almost on the poll's eve, the funeral of the iconic *fadista* Amália Rodrigues. Although little firm evidence of the overall health of a West European democracy can be gleaned from elections to the European Parliament, in June 1999 turn out in Portugal was up from 35.5 per cent in 1994 to 40.3 per cent. Whilst this percentage was impressive by British, Dutch and Finnish standards, the voters in 11 other EU countries put in better performances.[6]

Table 8.3 Turnout in Elections to the Assembly of the Republic, 1975-99 (% of registered electorate)

Year	1975	1976	1979	1980	1983	1985	1987	1991	1995	1999
% turnout	91.7	85.6	87.1	83.9	77.8	74.2	71.6	68.2	66.3	61.1

Source: STAPE, (various years)

The text of the fourth revision was designed to enable measures aimed at bringing disillusioned electors and politicians closer together. Article 109 records that 'the direct and active participation of men and women in political life is a fundamental precondition and instrument for the consolidation of the democratic system' and piously looks to a better balance of the sexes among political office-holders. A lengthy Article 115 brought innovatory prominence to the national referendum, which could be held on a wide range of issues with the approval of the President of the Republic. The result would only be binding when more than half of the registered electorate voted. Portuguese resident abroad could vote in a referendum, as they now could in presidential as well as legislative elections. Article 149 enabled electoral reform; although proportional representation remained, the possibility of single-member as well as multi-member electoral districts was mooted.[7]

The possibility of reform of the electoral system has been discussed with varying degrees of seriousness and urgency since the 1970s. The advantage of the system created in 1974 for elections to the Constituent Assembly, and restated in the Electoral Law of 16 May 1979, lay in the fairly close relationship between the number of votes cast for a major party or coalition, and the number of parliamentary seats awarded. However the two emigrant circles were anomalous. In 1999 their four deputies were elected by 43,040 electors (one Deputy per 10,760 votes on average) whereas the 226 from the mainland and islands were elected by 5,363,906 voters (one Deputy per 23,734 votes on average).

The perceived disadvantage lay in the way the closed-list system, with only party lists allowed, led to the selection of candidates in accordance with the desires of party headquarters in Lisbon. Parties therefore ruled the roost and by the late 1980s complaints were widespread, as in Italy, of this *partidocracia*. The election of candidates from outside the electoral district, or associated only with a part of it in large multi-member districts, meant that Deputies' links with those who voted for them were tenuous, when not non-existent. Consequently most electors knew little or nothing about the Deputies they elected. Over the years alternative systems were suggested, many in part influenced by the Federal German system in which the voter casts two votes, one for a candidate in a single-member constituency and one for a regional party list.

Electoral reform came to the fore in 1990 when Cavaco Silva's PSD, in an attempt to improve the efficiency and image of the Assembly, reduced the number of Deputies from 250 to 230. The smaller of the four main parties, the CDS and PCP, were particularly nervous of change along the lines of smaller electoral districts complemented by a national constituency, as they feared losing out. Both the PS and PSD were sufficiently interested to agree to set up a parliamentary commission for reform of the electoral law in 1992, but uncertainty as to the best system and an eye to party advantage led to failure (see Braga da Cruz, 1998). The closest reform has since come to success was in April 1998 when a PS draft, which would have created a mixed system of 99 single-member constituencies with larger districts and a national constituency, fell foul of the PSD's insistence on reducing the number of Deputies to 184 (Diário de Notícias, 1998). Plans were already afoot to reopen the question before the elections of October 1999 but it still remains to be seen whether, from the standpoint of party advantage, the PS and PSD will ever agree to take this leap in the dark in an effort to bring electors and elected closer together. As electoral reform needs to be approved by two-thirds of the Assembly of the Republic, these two political forces still have dual control over the key.

Apart from electoral reform to create a more direct link between elector and elected, the political class hoped that the practice of referendums would involve large numbers of voters in the political process. Two national referendums have been tried. The first, in June 1998, resulted from the narrowest approval in the Assembly for a PS proposal to allow the decriminalisation of abortion.[8] In the subsequent debate around the referendum question 'Do you agree with the decriminalisation of the voluntary interruption [sic] of pregnancy by the woman's choice, if carried out in the first ten weeks in a legally-authorised health institution?', the PS Prime Minister, António Guterres, and the leaders of the PSD and CDS-PP, all practising Catholics, campaigned for a negative answer. Whether because of the phrasing of the question, or the non-political nature of the issue, or the prevalence of good weather on the day of the poll, only 31.9 per cent of the electorate turned out. Thus the result of 50.9 per cent against decriminalisation was not mandatory, though the measure was dropped with the Minister of Justice announcing that in practice there would be no prosecutions. Though this referendum, pressed for by the opposition to show up divisions in the government and PS between Catholics and laicists, failed to obtain the participation of two electors out of three, its results demonstrated the continuing religious division of the country; between the North and the Islands, which voted against, and Lisbon and the South where the majority of voters were in favour of the proposition.

The second referendum was held in November 1998, with two

questions on regionalisation: 'Do you agree with the establishment of administrative regions?' and (though not for voters in the Islands) 'Are you in agreement with the creation of an administrative region in your electoral district?'. Again the opposition was keen on a referendum to play up divisions in the PS camp. The creation of eight regions on the mainland, although there were already five planning regions, led to fierce controversy. The main opposition party, the PSD led by Marcelo Rebelo de Sousa, was in principle in favour of regionalisation, but it was able to come out against this particular version of it. Of the 48.3 per cent of the electorate who turned out, just under 64 per cent answered 'no' to both questions, with a slender majority in favour only in the Alentejo, where the PCP is traditionally strong (see Corkill, 1999 and Silva, this volume, for a fuller discussion).

These experiences illustrate that the referendum has, thus far, not proved an easy way to obtain greater voter participation. Although electoral reform is set to return to the parliamentary agenda, this is unlikely to prove the panacea for the cynicism and alienation of voters. Agreement to have an uneven total number of deputies would, however, avoid awkward ties like that of October 1999.

The Evolution of the Major Political Parties

Since 1976 Portuguese political life has been characterised by a four-party system in which, particularly since the late 1980s, two parties (PS and PSD) have been dominant (see Table 8.1). There was a brief interlude in the mid-1980s when a fifth party, the PRD (Democratic Renewal Party – *Partido Renovador Democrático*), consisting of supporters of the austere Ramalho Eanes, hoped to establish itself as a major centrist contender among and against the other four main parties. However, the PRD's popularity as a protest party in 1985 soon faded. Its voters, many disillusioned with the PS of Soares because of his leadership and the austerity measures he put through with the PSD in 1983-85, for the most part transferred their allegiance to another austere leader, Cavaco Silva of the PSD.

The life of political parties is governed by the decree-law of November 1974 on their legalisation. Financial and campaigning aspects of their activities are now regulated by a law of 1998, which continues state funding in proportion to electoral success and allows private funding.[9] Critics of state funding are answered with the claim that sole dependence on private funding would lead to increased levels of corruption. Although limits are put on electoral expenditure, creative accounting is generally believed to overcome these restraints. Voters' suspicions of sharp practice are reinforced by the possibility of Deputies stepping down at will and being replaced by unelected candidates who were next on their party's list.

The basic pattern of party politics was set by the elections to the Constituent Assembly on 25 April 1975 when the PS and the PPD (People's Democratic Party – *Partido Popular Democrático* which in 1976 changed its name to PSD), emerged as the two main parties (see Table 8.1). From 1983 the need for the main parties to win over centrist floating voters became more obvious, as did the accompanying tendency to vote tactically to keep out the main political enemy. As a result the PCP and its coalition satellites on the left, and the CDS on the right, went into decline; the latter more rapidly since the line between PSD and CDS, in national coalition except in the Islands from 1979 to 1983, was more fluid than that between the PCP and the PS. The PS has since the conflicts of 1975 refused to enter into coalition with the PCP at national level, though co-operation in Lisbon municipal politics began in 1989.

The effects of change in hegemony from Marxism to neo-liberalism have even found an echo in the PCP, initially founded in 1921. Under the leadership of Álvaro Cunhal until 1992, the party adhered to an intransigent pro-Moscow line. Ghettoised for its activities in 1975, its role, and that of its satellite partners in electoral coalition and of the Intersindical trade union it dominated, became that of defender of the socialist gains of 1974-75 against the creeping capitalist counter-revolution represented by governments constrained by national and global economic circumstances (Patrício and Stoleroff, 1997). Although the shadow of the veteran Cunhal continues to hang over his successor Carlos Carvalhas, the party reset its goal, at least tactically, as the attainment of an 'advanced democracy' rather than Soviet-style Marxist-Leninist socialism (Cunha, 1997). In the late 1980s and early 1990s it lost a number of frustrated prominent personalities who had wanted reform, while its membership aged and its electoral support, centred in the old industrial areas of Lisbon and Setúbal and the heartlands of agrarian reform in the Alentejo, slowly shrank. Nevertheless the PCP led coalition's share of the vote increased in the low poll of October 1999 from 8.6 to 9.0 per cent, giving it two more seats in spite of losing some 20,000 votes.

In 1999 the PCP led coalition was challenged by a new *Bloco de Esquerda* (BE – Left Bloc), which won two seats in Lisbon. BE is essentially a 'new left' coalition of former Maoists from the UDP (*União Democrática Popular* – People's Democratic Union), Trotskyists from the PSR

(*Partido Socialista Revolucionário*) and former members of the old MDP front (*Movimento Democrático Português* – Portuguese Democratic Movement) who formed the Século XXI group after breaking with the PCP. The coalition's aspiration is to challenge ultra-liberal capitalist globalisation in the interests of greater social justice and equality. Critical of the PCP's failure to resist the trend, BE's 'democracy for socialism' looks to full employment, shorter working hours, more recycling and human rights for all (Bloco de Esquerda, 1999).

On the right, the CDS was founded in July 1974 with the encouragement of MFA officers anxious to cover the right flank of the PPD (Freitas do Amaral, 1995). Under the leadership of Diogo Freitas do Amaral it behaved as a conservative Christian-Democrat party, which voted against the socialist constitutional text of 1976 and took part in the AD coalition with the PSD in an effort to end leftist hegemony. Contrary to its intentions the CDS emerged weaker from the AD experience and went into decline. New leaders and policy changes failed to halt the process of subordination to the PSD. First came Lucas Pires and a more liberal stance, then a return to conservative Christian Democratic values under the former Minister of Salazar, Adriano Moreira, then an unsuccessful attempted repositioning in the centre with the return of Freitas do Amaral. In 1992 the young Manuel Monteiro became leader and 'refounded' the party the following year with the name *Partido Popular* (PP – Popular Party). His Euroscepticism lost the party its membership of the European People's Party (it allied at European level with the Gaullists and Fianna Fáil) but failed significantly to revive its fortunes (Robinson, 1996). Monteiro was replaced in March 1998 by his former chief advocate Paulo Portas who, with the failure of a coalition strategy with the PSD, was in 1999 left in an unenviable position; declining traditional rural support in the North and the lack of a secure foothold among the new urban middle class. In the low turn-out of the October election the PP's share of the vote fell from 9.1 to 8.3 per cent, but it kept its 15 seats despite a loss of some 80,000 votes.

Of the two major parties, the PS has probably changed most ideologically. Founded in West Germany in 1973, the party had very few members before the coup of 1974 but then expanded rapidly under the leadership of Mário Soares as the main opponent of the PCP, to achieve electoral success in 1975. In government the rhetoric had to give way to the exigencies of economic policy in a capitalist world. Soares's promise not to 'put socialism in the drawer' in 1977 wore increasingly thin. By the mid-1980s he had persuaded the party to change its programmatic orientation from Marxism to a democratic socialism defined as 'the deepening of political democracy with a view to broadening it out into society as a whole' (Robinson, 1991; Sablonsky, 1997). Lack of electoral success and the activities of Soares's adherents within the PS made for lack of stable leadership after his resignation in 1985. First Almeida Santos led the unpopular party to its worst result in the elections of October 1985, then the economist Vítor Constâncio failed to quell rebelliousness in the party, resigning in October 1988, and his successor Jorge Sampaio paid the price for electoral defeat in 1991. In 1992 António Guterres became the Catholic leader of a mostly laicist party and attracted support from outside the PS, which put him in good stead when the electorate expressed its weariness with the PSD in 1995. He subsequently remained a popular leader of an administration which continued the privatisation policy of Cavaco Silva and governed pragmatically; including in its ranks ex-Communists, a former parliamentary leader of the PSD and, for two years, a former Minister in Marcello Caetano's pre-1974 government. As expected, the PS again received most votes in the October 1999 elections, but its failure to achieve its objective of an absolute majority of seats detracted from its victory, based as always on support from across the country.

The PSD (originally PPD) was founded in May 1974 as a social democratic party which initially sought to join the Socialist International (which the PS opposed). Its early years were marked by the turbulent but successful leadership of Francisco de Sá Carneiro, who repositioned it to pick up right-wing voters at the expense of breaking with most of the parliamentary party of 1976 (including Sousa Franco, Guterres's Finance Minister from 1995 to 1999). Sá Carneiro formed the AD with the CDS and a small monarchist party and, although he was killed in an air-crash in December 1980, the AD experience put the party in power (shared or alone) for 16 years, giving it powers of patronage which critics dubbed 'the Orange State' (after the identifying colour of the party). Pinto Balsemão succeeded him but was entangled in the internal politics of the AD coalition. Mota Pinto, leaning to the left, kept the PSD in power with the PS until his death in 1985. Later that year Aníbal Cavaco Silva, leaning to the right, took over and, in improving economic circumstances, consolidated Portuguese democracy on a neo-liberal basis (Frain, 1998). His successor Franco Nogueira lost the 1995 election and it was left to Marcelo Rebelo de Sousa to try the expedient of a new AD (*Alternativa Democrática*) coalition with the CDS-PP. It was never popular in the party and, untested, it fell apart in March 1999. Rebelo de Sousa was succeeded by a contender with a rightist image, Durão

Barroso, who picked a member of the party's left, Pacheco Pereira, to head the European election list: both had been Maoists in 1974-75, though of different persuasions. With little time to find a distinctive strategy and image, Durão Barroso predictably failed to defeat the PS in October 1999. The party's bedrock of support remains the North and the Islands, although the PS has now stolen a march on it in the Azores.

In 1974-75 parties adopted stances and rhetoric to the left of what could be termed their natural position so as to retain credibility in the revolutionary process. Since then a sea change of ideas in Western Europe has taken place under the pressure of global economic forces. Portugal and its parties have shared in this sea change. While even the PCP now talks of 'advanced democracy', the PS has jettisoned its Marxism and now governs like 'New Labour' in Britain. The PSD no longer seeks to join the Socialist International. After a period of alliance with Liberal parties, it joined the European People's Party in 1996.

Conclusions

Portugal has changed economically, socially and politically in the past quarter century. In politics, the image of instability and uncertain democracy of the 1970s has given way to stability. There have been only two Presidents of the Republic since 1986 (Soares and Sampaio, elected in 1996) and only two Prime Ministers since 1985 (Cavaco Silva and Guterres). A full member of the EU since 1986, signatory of the treaties of Maastricht and Amsterdam and in the first wave of countries to adopt the Euro, Portugal has become a fully fledged and consolidated European democracy (Magone, 1997).

The problems Portuguese democracy faces are now the problems faced by other, longer established democracies, which is evidence of the degree of political convergence achieved by Portugal in 25 years. Within this picture of success there are inevitably areas which are less than perfect. As President Sampaio pointed out in his speech commemorating the 25th anniversary of the revolution of 25 April 1974, there were in 1999 'signs of growing distance in the relationship between the citizen and the system of representation', indicators of which were the rates of electoral abstention and 'the absence of volunteers to help with the business of electoral scrutiny and monitoring, which even ten years ago mobilised thousands of party militants' (Diário de Notícias, 1999). These fears were confirmed by the low turn-out for the Assembly elections on 10 October 1999.

At least in part, as a columnist in the *Diário de Notícias* argued the day following the President's speech, this may be connected with social and technological changes which have brought an increasingly media-dominated society. There has emerged a 'neo-elitism' of media leaders with media advisers. Spin doctors have become more important than mass membership or party structures as decisions are made by fewer and fewer individuals. The columnist concluded that there is a widening gap between the conventional political party with its inherited structures and the new forms of the exercise of power. This gap constitutes a challenge to the system of popular representation, while there is a danger that politics are mistaken for their media image (Santos, 1999). These are problems Portugal shares with other, more advanced democracies in western Europe, where indices of participation are also sliding as electors sense that changing a government may make little difference. As the integration process advances in the EU, there is inevitably less real choice to offer national electorates. For Portugal, as for its EU partners, the challenge of the new century lies in how to nurture increased participation in the formal democratic process by a well-informed and responsible electorate.

Notes

1   Article 174 of the Constitution of 1976 set the duration of the Assembly of the Republic at four years; in the case of premature elections having to be called, these elections would not be defined as starting a new legislative quadriennium. Hence elections again in 1980 although the AD coalition had a working majority (see Table 8.1).

2   The PS's Marxism, it was explained in the party's principles, was not Stalinist: its Marxism was 'perpetually rethought' (PS, 1975).

3   Two PS Deputies were elected for 'Europe' by 26.5 per cent of the registered electorate, and one PS and one PSD Deputy were elected for 'Outside Europe' by 23.3 per cent of those registered. Results for the emigrant electoral circles were published at http://www.stape.pt (23 October 1999) and those for the Mainland and Autonomous Regions at http://resultados.cne.pt (20 October 1999).

4   For the text of an Azorean manifesto of independence, produced by FLA (Azorean Liberation Front) in 1975, see *Diário de Notícias (edição Internet)*, 6 June 1999 at http://www.dn.pt

5   In 1986 Mário Soares was elected the first civilian President for sixty years.

6   In the UK turn out was 24.0 per cent, in the Netherlands 29.9 per cent and in Finland 30.1 per cent; immediately ahead of Portugal were Germany with 45.2 per cent and France with 47.0 per cent. The EU overall turn out was 49.4 per cent, see http://www2.europarl.eu.int/elections.

7   The text of all 299 articles of this currently valid version are available with an English translation at http://www.parlamento.pt/frames/revisao_constitucional

8   A law of 1984 had allowed abortion in cases of rape, where there was a risk to the
       mother's health or in cases of malformation of the foetus.
9   The full text of the Law on Financing Political Parties and Electoral Campaigns of 18
       August 1998 is available at http://www.parlamento.pt

References

Bloco de Esquerda (1999), *Projecto de Declaração: Começar de Novo*, BE, Lisbon.

Braga da Cruz, M. (ed) (1998), *Sistema Eleitoral Português: Debate Político e Parlamentar*,
    Imprensa Nacional, Casa de Moeda, Lisbon.

Corkill, D. (1999), 'Portugal's 1998 Referendums', in *West European Politics*, vol.22, no.2,
    pp.186-192.

Cunha, C. (1997), 'The Portuguese Communist Party' in T. Bruneau (ed), *Political Parties
    and Democracy in Portugal*, Westview Press, Boulder, CO, pp. 23-54.

Diário de Notícias (edição Internet) (1998), 'Reforma do Sistema Eleitoral Chumbado', 24
    April, at http://www.dn.pt

Diário de Notícias (edição Internet) (1999), 26 April, at http://www.dn.pt

Frain, M. (1998), *PPD/PSD e a Consolidação do Regime Democrático*, Editorial Notícias,
    Lisbon.

Freitas do Amaral, D. (1995), *O Antigo Regime e a Revolução: Memórias Políticas 1941-
    1975*, Bertrand/Nomen, Venda Nova, pp.165-98.

Magone, J.M. (1997), *European Portugal: The Difficult Road to Sustainable Democracy*,
    Macmillan, Basingstoke.

Manuel, P.C. (1995), *Uncertain Outcome: The Politics of the Portuguese Transition to
    Democracy*, University Press of America, Lanham, MD.

Partido Socialista (PS) (1975), *Declaração de Princípios, Programa e Estatutos do Partido
    Socialista, aprovado no Congresso do PS em Dezembro de 1974*, PS, Lisbon.

Patrício, M.T. and Stoleroff, A (1993), 'The Portuguese Communist Party: loyalty to the
    "Communist ideal"' in D.S. Bell (ed.), *Western European Communists and the
    Collapse of Communism*, Berg, Oxford, pp.69-85.

Robinson, R.A.H (1979), *Contemporary Portugal: A History*, Allen and Unwin, London.

Robinson, R.A.H. (1991), 'The Evolution of the Portuguese Socialist Party, 1973-1986, in
    International Perspective', *Portuguese Studies Review*, vol.1, no.2, pp.6-26.

Robinson, R.A.H. (1996), 'Do CDS ao CDS-PP: o Partido do Centro Democrático Social e o
    Seu Papel na Política Portuguesa', *Análise Social*, vol.31, no.138, pp. 951-73.

Rogaly, J. (1999), *Financial Times*, 24 April.

Sablonsky, J.A. (1997), 'The Portuguese Socialist Party' in T. Bruneau (ed), *Political Parties
    and Democracy in Portugal*, Westview Press, Boulder, CO, pp.55-76.

Santos, J de Almeida (1999), 'O Partido Mediático' in *Diário de Notícias (edição Internet)*,
    26 April 1999 at http://www.dn.pt

STAPE (Secretariado Técnico para os Assuntos do Processo Eleitoral) (various years) *Results
    of Elections to the Assembly of the Republic*, STAPE, Lisbon.

9 Local and Regional Government: Continuity and Innovation in Local Governance

CARLOS NUNES SILVA

Introduction

Portugal has experienced profound economic, social and political change across the last 25 years (Santos, 1993, Reis, 1994). In the area of local public administration these major changes are evident across all dimensions of its operation. The nature of local government with respect to its structure, competencies, financing, planning instruments and production of public services has developed significantly, and its changing role and relationships with other public, private and voluntary sector actors, has led to the emergence of new forms of local governance. However, against this background of change, continuity is also evident. This is particularly notable with respect to the continuing low proportion of total public administration expenditure accounted for by local government, and the absence of a regional level of self-government (Silva, 1995a, 1996, 1998a; C. Oliveira, 1996; Syrett and Silva, 1998).

With the establishment of the new political regime in the 1974-76 period, the Portuguese local political system underwent a fundamental shift from local administration to local self-government, characterised by directly elected bodies with political, administrative and financial autonomy. A new range of competencies were transferred from central government, a process still in progress, and a new finance system was introduced, guaranteeing a significant level of autonomy to municipalities and parishes. This general trend towards increased decentralisation took place within a changed wider context of increased integration into the European economy, as well as significant technological, social and political changes. These changes provided a stimulus for the emergence of new forms of service production and local governance, albeit in a geographically uneven pattern.

However, despite the overall improvement in the position of local government in Portugal, there are still major reforms to be carried out both with respect to local government organisation and the local finance system.

This chapter will examine the present structure of local and regional government and the prospects for local governance in the near future. Specifically, the chapter addresses three key questions. What are the main organisational principles and structures in local government? What are the consequences of the missing regional tier? What is the future prospect for local government and its role in a broadened base of local governance? The chapter begins with an examination of the structure of local government and the processes that have led to change. The second section moves on to discuss the absence of a regional tier, and the final section explores the future prospects for local and regional government in Portugal and the likely impact on the urban governance base.

Local Government in Portugal

Local Government Structure

The 1974 military coup put an end to the authoritarian regime and to the corporative organisation of the state, including its particular structure of local administration and politics. For the first time in half a century local councillors were democratically elected and a new system of sub-national government was put in place. The Portuguese constitution of 1976 defined three tiers of local government: parishes (4,241), municipalities (308) and administrative regions.[1] The last of these has never been implemented and

Table 9.1 Population by Municipality and Parish (average)

		Total	Mainland	Azores	Maderia
1974	Municipality	28, 889	30,135	14,521	22,664
	Parish	2, 182	2,154	1,985	4,704
1998	Municipality	32,329	34,008	12,832	23, 573
	Parish	2,348	2,342	1,625	4,802

Source: INE, Recenseamento da População

their formation was rejected in a national referendum held in 1998. Government for the islands of the Azores and Madeira is via two 'autonomous regions', a form of political autonomy which is not considered as part of the local government system, but rather as a form of political decentralisation (see Figure 9.1).

Municipalities are grouped together in 66 officially registered municipal associations as well as in 52 groups which are supported by technical offices (*Gabinetes de Apoio Técnico* (GATs)). The average population size of local government units was 32,329 in 1998 (see Table 9.1 and Figure 9.2) making them among the largest in Europe. As a result, the units are big enough to sustain the cost of the increasingly professionalised administration which has emerged over the last 25 years. Local government is based on the constitutional right to handle a number of tasks within their geographical area, to an extent laid down by Parliament, and subject to state supervision on legal matters.[2]

During the period of the *Estado Novo* (1926/33-74) local government was strictly controlled by the central administration, with the Mayor being appointed by central government and the Mayor then appointing the executive. After 1976 all members became directly elected, initially for three years, but since 1985, for four year terms. Participation in elections to the municipality is via a party list system,[3] although at the parish level it is possible for groups of independent citizens to participate. Every level of

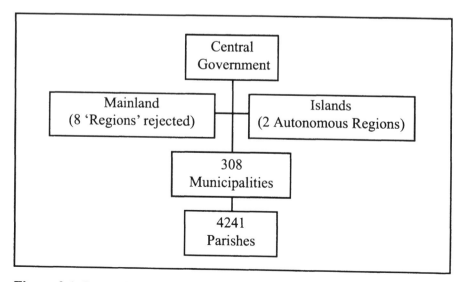

Figure 9.1 Local Government Structure, 1999

local government (municipalities and parishes) has two boards, one executive and another deliberative.[4] In the case of municipalities both are elected
directly by proportional representation.[5] The number of members varies
from 5 to 17 in the executive board (*Câmara Municipal*), according to the
number of electors. The deliberative board, the municipal assembly
(*Assembleia Municipal*), is composed of directly elected members and non-
elected ones (i.e. all the presidents of the parish executives).

The head of the *Câmara Municipal* (the Mayor or *President*) is the top
of the list that gets most votes and a key figure in local government and
politics. The *Câmara Municipal* is served by an administration comprising
several departments (for example covering administration and finance,
planning, physical infrastructures, education, culture and leisure, etc.). Local
authorities have considerable room for manoeuvre in the organisation of
their administration although they must follow certain general rules
established in law.[6] As the head of the local administration the Mayor must
ensure that decisions by the local executive and deliberative councils are
executed. This role is shared by some, or all, of the other members of the
executive council (*vereadores*).

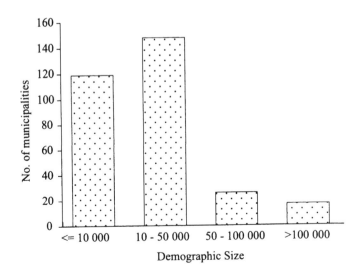

Source: INE, Recenseamento da População

Figure 9.2 Municipalities by Population Size, 1998

Elections and Party Politics

Despite the close association with the municipal population and the growing politicisation of the Mayor's role, participation in local elections is lower than in national elections (Figure 9.3). This low level of mobilisation in local elections is partly due to the inadequacies of the centralised and rigid party system for local politics. In addition, the different role played by the media in the two types of elections, and the reduced volume of resources used in local elections campaigns compared to national ones, are also important factors.

Comparison between the electoral results at the local level and those for the national Parliament indicate certain specificities relating to municipal elections. First there is a stronger degree of local consensus rooted within personal relations and a pragmatic assessment of the capabilities of different administrations. Second, there is a strong personification of the vote associated with the particular character and image of the Mayor. Finally, voting in local elections is frequently used as a means to protest against national policies. There are also clear geographical variations in local elections, associated with factors such as the socio-economic context,

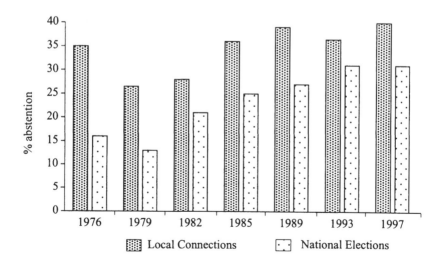

Source: STAPE, Local Election Results

Figure 9.3 Voting Turnout in Local and National Elections, 1976-97

religious practices and the proportion of the vote accounted for by urban, rural and industrial populations. Local election results display marked differences between the North (favouring the political Right) and the South (favouring the political Left), although such differentiation is declining.

Elected politicians in the four main political parties (conservative (CDS/PP), social-democrat (PSD), socialist (PS) and communist (PCP)) exhibit significant differentiation in terms of socio-economic status, gender[7] and age (STAPE, 1987; 1993). There are also identifiable differences in local government policies related to the nature of party political control. Political parties from the Left and Right accounted for an equal number of municipal presidencies in the first local elections of 1976 (152 each), whilst in subsequent elections right wing parties had a majority of presidencies in 1979, 1982, 1985 and 1993, and left wing parties had majorities in 1989 and 1997 (Figure 9.4). Control by the Left is associated with higher levels of spending across several expenditure categories and proportionally higher spending than municipalities controlled by the Right. The empirical evidence indicates that in municipalities which remain governed by the same party in successive terms, left wing controlled municipalities tend to have a higher expenditure growth than those run by conservative parties. Where municipalities

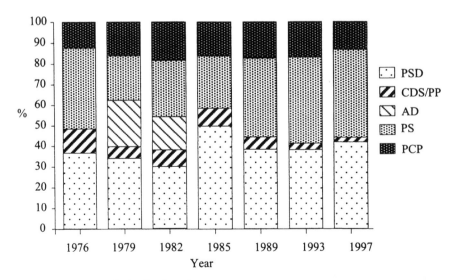

Source: STAPE, Local Election Results

Figure 9.4 Percentage of Presidencies of Municipalities (Mayors) by Political Party, in all Democratic Local Elections, 1976-97

have changed political majorities in two succesive elections, those that moved from right to left wing control experienced higher average spending increases than those that moved from left to right. Party political differences can also be identified in the areas of housing policy, urban planning, social and cultural policy (Silva, 1995a; 1995b; Pauleta, 1999).

In the local elections of December 1997 it was possible for the first time for a foreign citizen, satisfying certain eligibility criteria, to vote and to be elected. In total, 14,516 citizens from 21 countries registered as electors. Of those, 10,482 (72.2%) were from Cape Verde and 3,171 from EU countries. Around 68% (9,922) lived in the Lisbon Metropolitan Area (AML). Whilst the number of elected foreigners was small and limited to secondary posts, their presence was greatest in left wing parties, mainly in suburban municipalities in the AML.

Competencies and Finance

Local government expenditure focuses on the provision of local public services associated with urban planning, basic infrastructures and urban

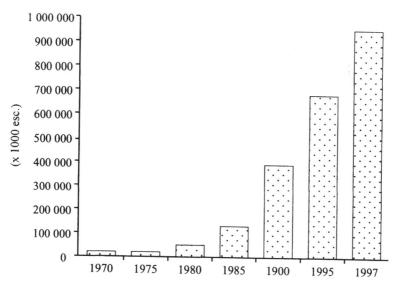

Source: INE, *Estatísticas das Administrações Públicas* and *Estatísticas das Finanças Públicas*

Figure 9.5 Evolution of Municipal Revenue, 1970-97

services. Social welfare and economic development remain secondary functions compared to spending in these areas by central government (Silva, 1995a; 1998b). The local finance law published in 1998 changed important aspects of the previous 1987 legislation.[8] Whilst broadly welcomed as a positive set of reforms by the National Association of Portuguese Municipalities (ANMP), the legislation also provoked dissatisfaction; particularly concerning the total amount of resources allocated to local government, as well as from smaller municipalities who feared they would be relatively worse off. Under this legislation the electoral promise made by the socialist party to double grants to local government across a four year period remained unrealised. In fact, given the rate of increase across the 1996-1998 period, it would take 14 years to meet this commitment. However, despite the slow rate of change, the longer term trend is towards a significant increase in total revenues to local government since 1974 (Figure 9.5).

Municipal revenue currently has three main components: state budget transfers (comprising two block grants plus conditional grants), tax revenues, and loans. Additional revenue comes from fees and other minor items. Within the local finance system the role of local taxation has been

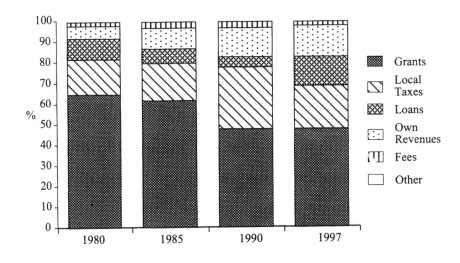

Source: INE, *Estatísticas das Administrações Públicas* and *Estatísticas das Finanças Públicas*

Figure 9.6 Structure of Municipal Revenue, 1980-97

increasing relative to grants from central government (Figure 9.6); a major change to what existed before 1974 when there was almost total dependency on revenue from central government. The two block grants comprise the *Municipal General Fund* (FGM),[9] which provides the majority of finance for municipalities to fulfil their basic duties, and the *Municipal Cohesion Fund* (FCM), which seeks to equalise tax revenues between municipalities on a per capita basis. Together these grants seek to promote greater equity between the state and local authorities, and between municipalities themselves. In terms of local taxation there are three major municipal taxes: *Sisa* (a tax on the acquisition of immovable property) which represents 8.5% of total municipal revenue; *Contribuição Autárquica* (a tax on immovable property) representing 7.5%; and a surcharge on corporation income tax (*derrama*) 4%. Loans represent a minor proportion of municipal income (7%) and user charges and fees account for only around 3%.

The organisation of municipal services has undergone profound changes since the early 1990s. However, contrary to what has happened in many other EU countries, these changes have not been a simple consequence of adapting to reduced levels of public funding. In the Portuguese context, organisational changes have resulted from a combination of ideological factors and pragmatic concerns related to the need to improve the quality and efficiency of service delivery. This resulted in the existence of a cross party consensus relating to the need to introduce new forms of management at the local government level. From the early 1990s a mixture of management innovations emerged which have spread across most municipalities. These included increased contracting-out of services, use of contract management, creation of municipal public enterprises,[10] use of new information technology in all services, and of leasing in different types of operations. Regular use of contracts with private enterprises to provide municipal services (e.g. in education, transport, sewage, infrastructures, road building and maintenance, cultural activities, industrial parks, markets, etc.) and an increase in public-private partnerships, were also part of this innovation process across most policy areas. As a result of these changes, there has been a significant increase in the role of the private sector in the municipal sphere, and local government has been able to pursue its service provision role with much greater flexibility.

Since 1974, local government within Portugal has undergone a fundamental transition from being an arm of the central state to a well established system of local self-government. This transition has comprised a series of major changes: from political dependency to local political autonomy; from financial dependency to fiscal autonomy; from specific competencies to overall responsibility; from corporative representation to forms of democratic election via proportional representation; from local elections exclusive to national citizens, to local elections open to foreign residents. Yet, despite these changes, local government (including both municipalities and parishes) remains responsible for less than 10% of total public administration expenditure; a figure which puts Portugal at the bottom of the EU scale. This high level of hierarchical dependency marks a clear continuity in the nature of local government between the two political regimes. The other major element of continuity is the lack of a regional level of self-government, and this will be considered in the following section.

'Administrative Regions': the Missing Tier of Local Government

The regional question has been considered in a number of different ways in Portugal across the twentieth century (Caetano, 1982; Santos, 1985; L. V. Oliveira, 1996; 1997; Barreto, 1998; Silva, 1996). During the *Estado Novo* the regional level was seen as one of several levels in the corporative structure, with boards elected by corporations and also with members nominated by central government. The 1933 Constitution recognised 11 regions (then called Provinces). These were weak forms of sub-national administration characterised by a lack of political legitimacy, few competencies in only a limited number of sectors, limited fiscal resources, and strict central government control of their actions. In the first constitutional revision of 1959, the Provinces were abolished and replaced by 18 districts (*distritos*) which provided a regional level of administration, but again, one lacking any significant relevance. However, in the last period of the dictatorship several attempts were made to strengthen the regional level and implement regional planning. These ideas were advanced within the regime by those promoting greater economic liberalisation and were based mainly on technocratic arguments.

It was therefore no surprise that immediately after the military coup in 1974, that the regional question moved centre-stage on the political agenda. Indeed, with the replacement of the authoritarian regime, the introduction of a regional administrative level was seen as a key factor in the process of democratisation; one positively valued by all political parties and sectors of civil society (Caetano, 1982; Santos, 1985; Barreto, 1998). In the ensuing 1976 Constitution, administrative regions were identified as the third tier of local government. Since then the regional issue has been discussed by every political party in all electoral campaigns, yet with nothing ever implemented once a party was elected to government.[11]

Integration into the European Community (EC) in 1986 provided renewed impetus for the development of an effective regional administration within Portugal. As a result, in 1991, a law creating Administrative Regions was approved by Parliament with cross party support.[12] However, in 1995 the situation changed dramatically when the PSD, then the party with an overall majority in Parliament, announced that it was no longer in favour of administrative regions. Other political sectors followed suit and a national campaign started against the perceived threat of regionalisation to national unity. In response, the socialist party made the creation of regions a central element within their manifesto for the 1995 election, a position also supported by the communist party.

Following the election of the PS government a new legislative initiative for the creation of regions was launched, including proposals mapping out the regions. The regional map finally adopted comprised 8 regions and reflected the PS's original proposals with minor modifications imposed by the PCP (Table 9.2). As part of their opposition to the proposals, the PSD forced a referendum on the issue, arguing that without a clear 'yes' vote it would vote in Parliament against the implementation of the planned region-alisation. The eventual referendum in 1998 resulted in a clear rejection of the proposed administrative regions, although the turnout of 48.3% was very low.

There are many factors which contributed to the 'no' vote.[13] However, the four most important arguments against administrative regions related to the perceived risk to national unity, the cost to the ordinary taxpayer, the increase in bureaucracy, and the proposed regional map; particularly with respect to concerns over regional boundaries and the fact that regional capitals were not defined (Barreto, 1998). Furthermore, as the low turnout indicates, the referendum proposals failed to capture the interest of the vast majority of the Portuguese people, reflecting the marginal importance many felt towards the whole regionalisation process.

As a result of the referendum, it appears that there will be no development of administrative regions for many years to come, despite the fact that they continue to exist as a tier of local government within the political constitution. In response, different political parties have advanced new proposals dealing with particular aspects connected to the regional issue. These relate to the de-concentration of central government services and the harmonisation of their boundaries limits,[14] the reform of current

Table 9.2 Proposed 'Administrative Regions' for 1998 Referendum

Regions	Area (km²)	%	Population N°	%	Population Density	No. Municipalities	No. Electors
Entre Douro e Minho	7 624.9	8.6	2 714 712	29.0	356	43	2 291 979
Trás-os-Montes e Alto Douro	12 272.1	13.8	466 751	5.0	38	32	424 812
Beira Litoral	11 092.75	12.5	1 369 789	14.6	123	56	1 172 472
Beira Interior	11 474.14	12.9	386 693	4.1	34	23	344 588
Estremadura e Ribatejo	9 835.35	11.1	853 935	9.1	87	34	733 658
Lisboa e Setúbal	4 284 86	4.8	2 688 673	28.7	627	24	2 365 533
Alentejo	27 224.78	30.7	549 362	5.9	20	47	464 981
Algarve	4 988.49	5.6	341 404	3.6	68	16	304 312
Total (Mainland)	88 797.37	100.0	9 371 319	100.0	106	275	8 102 335

Source: INE, Recenseamento da População

forms of government in the major metropolitan areas of Lisbon and Porto, and the creation of new metropolitan institutions for other major urban areas.[15] In 1999, new legislation on municipal associations as well as, for the first time, for associations of parishes, was introduced.[16] In fact one consequence of the rejection of the regional proposals is that all political parties now agree that the reinforcement of municipalities is a national priority. In the absence of regional administrations, the need to re-define the roles of central and local government has become a major political issue, particularly in relation to the emergence of new forms of urban governance. A brief discussion of recent and likely future changes in the local government system, is the focus of the next section.

Key Issues and Future Prospects

In order to cope with the rapid socio-economic transition which Portugal has undergone in the late twentieth century, local government is faced with the need for fundamental modernisation relating to increased autonomy, powers (via greater subsidiarity) and flexibility in management, organisational structures and working methods. At their 1998 conference, the National Association of Portuguese Municipalities (ANMP) proposed a series of measures to achieve these aims. These included a reduction in bureaucracy, administrative rationalisation, the creation of municipal or mixed enterprises, the decentralisation of competencies and resources, new working practices for local politicians, a greater participation in the management of EU funds, and the reinforcement of municipal associations. However, currently there is no policy agenda to change radically the present structure and rationale of local government. In fact the policy emphasis appears more directed towards evolution rather than revolution. The introduction of a series of pieces of legislation in the late 1990s reflects recognition of the need for change and has created a new context for the operation of local government.[17] The section will consider some of these recent changes in greater detail and explore the major issues faced by Portuguese local government along a number of key dimensions.

Decentralisation and Partnership

The rejection of the proposals to create administrative regions has created a situation whereby the structures of sub-national governance require

fundamental reappraisal. This is reflected in the 'XIV Government Programme' (1999-2003) which took as its central concerns the territorial
organisation of the state, the reform of the state apparatus, the decentralisation
of new functions to the municipalities, municipal associations and
metropolitan authorities, and the territorial co-ordination of public policies.[18]
The absence of administrative regions has brought increasing pressure to
develop new functions and competencies at the municipal level as well as to
develop new and existing regional level institutions. There are proposals to
develop a range of new regional institutions (e.g. Regional Economic and
Social Councils, Regional Co-ordination Councils (for public administration
only), and Regional Development Agencies). Consideration is also being
given to proposals for the de-concentration of national state departments as
well as to reform of existing governmental regional organisations such as the
Governador Civil and the Regional Planning Commissions (CCRs).[19]

The creation of municipal and regional public enterprises has also
created new possibilities for expanding the municipal role, either acting
alone or in association with others.[20] Partnerships with private capital have
expanded in the context of local economic policies and the creation of
municipal enterprises or joint ventures with private capital have become
increasingly common.[21] There is also growing acceptance that more innovative
and entrepreneurial forms of organisation are required in order to improve
the delivery of local public services. For some services this may result in
'privatisation', in the sense of public concessions granted for 20-25 year
periods.[22]

Local Finance System

The Portuguese local financial system is highly centralised, whether
assessed in terms of the percentage of public expenditure it accounts for, or
its ability to raise local taxes (Silva, 1998a). Its future development therefore
has to confront four associated challenges: the high degree of centralisation
of public administration expenditure; the weak capacity of municipalities to
raise local revenues; the lack of a diversified local tax base to provide greater
revenue stability; and the equity of local taxes, notably the regressive
character of immovable property taxes and user charges. Within this context,
immediate reforms are required in two areas. First, a substantial increase in
the total volume of resources; an area where the PS government has
promised to act, but has not yet delivered.[23] Second, the introduction of
income tax as a local government revenue source. Under the last constitutional

revision, local government was given the ability to raise taxes, but how and whether these powers will be developed still remains uncertain.[24]

Competencies

Whilst in general local government is in favour of increased competencies in areas such as education, planning, environment, public security and social policy, increasingly it is seeking to ensure that appropriate levels of financial resources are also transferred to the local level.[25] This position draws strongly upon past experiences where competencies were transferred to local government in the areas of civil protection, conservation of school buildings, river conservation and road de-classification, but without adequate financing, thus making it impossible for local government to meet its responsibilities.

Recognition of the growing differences between municipalities in terms of their needs and capacities has led to a growing movement of local councillors advocating a variable transfer of competencies, related to municipality size and agreed on a voluntary basis. This position has now been accepted by Parliament in new legislation on municipal competencies.[26] This legislation will fundamentally change the uniform nature of the present system. How this legislation will be implemented remains unclear, however rather than a host of individual agreements between the state and each municipality, there is likely to be some definition of different categories of municipalities to which different levels of competencies will be decentralised.[27] As part or this overall process local authorities will be able to devolve greater responsibilities to the parish level. There are already numerous examples of voluntary decentralisation of competencies to parishes, which demonstrate the capacity of this lower level to carry out more functions in a more efficient manner, if given adequate resources (Silva, 1998b; Pauleta, 1999).

Structures and Administration

The existing electoral system for the *Câmara Municipal* (CM) creates a number of difficulties for Mayors in their daily management activities. Moves to improve this situation focus on proposed changes to the electoral system which would permit indirect election and would allow a uniform political composition in the CM. Such a move would also require the control powers of the *Assembleia Municipal* to be strengthened. Other suggested reforms focus on the need for more full-time councillors and better remuneration of technical staff in order to keep good quality professional staff. within the municipalities. In terms of local level participation, some examples of new forms of direct democracy and decision-making have begun to emerge. For example, in April 1999 the first local referendum promoted by a parish council took place and was quickly followed by three more on quite different issues (e.g. new car parking, a seaside road, a cultural centre).

The modernisation of the administration of local government is an area where the need for reform has cross party support. A series of measures are currently under consideration which aim to clarify the role of municipalities, to give them greater autonomy and to provide them with greater flexibility.[28] The publication of a Municipal Code is seen as an important element within this process as this would rationalise and organise the fragmented legislation currently regulating different aspects of local government in Portugal. With regard to municipal management a major challenge is to develop processes which generalise best practice and promote the introduction of innovatory practice.[29]

Territorial Coverage and Urban Governance

Given major shifts in population distribution and travel to work patterns, there is a need to look once again at the basic territorial structures of municipalities and parishes. There appears to be a strong case for the creation of a number of new municipalities and parishes (via a process of segmentation), as well as the elimination of others whose existence and sustainability appears no longer justifiable. In the absence of a regional level of administration, the need to improve the working conditions of municipal associations in terms of funding and staffing levels, appears of fundamental importance (Silva, 1996). Alongside the development of municipal associations, the development of parish associations also appears an important new means of operation. These associations can be created to manage public equipment, common services and competencies decentralised from the municipality.

For the major urban areas, there is growing political pressure to consider the creation of new forms of administration. Debate has centred on the possible implementation of special administrative forms for the main urban areas, as well as the reform of the present system of metropolitan government towards a system of directly elected boards. Whilst all major parties agree the need to reinforce the metropolitan authorities, there is divergence concerning what form this should take. These range from the official position of the PSD, which favours some form of municipal association, to the position of the Communist Party, which favours the creation of a metropolitan, self-government institution (*autarquia metropolitana*) with a directly elected metropolitan assembly.[30]

Key Policy Areas

Almost all areas of municipal policy have experienced significant change since 1974, and they each face their own specific challenges in the future:

Planning The local planning system is one where radical changes have been introduced over recent years and more are expected.[31] Increased regulation of the planning system is evident in a number of key aspects including procedures relating to development activity and land expropriation.[32] With regard to urban planning the key issue remains the availability of urban land. In 1995, the Socialist Party announced the creation of a special credit line for the acquisition of urban land by municipalities, however this has not been implemented. There are also potentially important proposals to place the granting of central government licenses in sectors such as forests, mining, industry, retail, and tourism in municipal hands, to ensure that they go ahead in accordance with the municipal master plan. Municipalities also seem likely to gain special planning powers for certain problems, such as slums and re-housing programs in the large urban areas.

Road infrastructures Road building and maintenance remains a major sector of municipal activity. Here the key policy issue relates to the declassification of national roads and their decentralisation to municipalities. The challenge here for municipalities is to ensure that the central state repair roads before transference to the municipal level, and that roads with national traffic are kept under central state control.

Environmental issues Municipalities are increasingly involved with environmental issues, either through the local planning system, through specific instruments outside the formal planning system (e.g. via Municipal Environmental Plans, Local Agenda 21, Environmental Audits etc.), or via their involvement in partnerships with central government in programmes aimed at reducing energy consumption,[33] and in the future, air quality.[34] New legislation on environmental impact assessments has given a more important role to municipalities in the development process. The quality of water for human consumption, a traditional responsibility of municipalities, has improved significantly over the last twenty years as a result of major investments in infrastructures.[35]

Social policy   In the area of social policy there is increasing recognition that local government actions are going beyond their legal obligations (Silva, 1998b; 1999; Pauleta, 1999). Consequently, there is a need to clarify local government and parish roles in this field and ensure greater co-ordination of local actions. Municipalities have been increasingly active as local partners in areas of employment creation and innovation, whilst education is a major area where new and increased municipal responsibilities are currently being developed.

Housing   Traditionally housing has been considered a central government responsibility with municipalities playing only a secondary role. Despite this, involvement in the provision of social housing is a traditional policy field for local government, and one which has witnessed increased activity in recent years within the metropolitan areas. However past proposals to transfer state social housing estates to municipalities remain largely unrealised.

Local economic development   Local economic development is another area of growing municipal intervention. The creation and modernisation of infrastructures as well as the development of a package of incentives to attract private investment, is already a common policy objective (Syrett, 1995). The creation of information centres for small and medium size enterprises (SMEs) is becoming increasingly common,[36] but the future of these types of interventions is largely reliant on the level of resources allocated to local government under the Third Community Support Framework (2000-2006). Direct involvement in productive activities, as part of a wider objective of strengthening local economic structure, is now generally accepted as part of the municipal role. In the future this role will also be pursued through municipal participation in the recently created Regional Development Agencies which now cover all of the country, and which aim to promote regional economic competitiveness (Syrett and Silva, 2001).

Conclusions

Portuguese local government has witnessed substantial growth and development over the last 25 years. Despite the failure to establish a regional tier, the increased role of local government in the regulatory process appears set to continue. However, in the near future, the local government system in Portugal also has to address four key problem areas. First, problems related to territorial structures, particularly the absence of a regional tier and the increased non-coincidence within metropolitan areas of places of work, political participation and use of public services. Second, problems of co-ordination associated with the complexity of the local government system and the emergence of new partners within urban governance. Third, financial problems related to the continued low levels of funding from central government and the limited local tax base. Fourth, problems related to declining levels of local level participation as evidenced by falling turn-outs in local elections.

Following the referendum vote against regionalisation, current trends favour the implementation of a reform process which will increase the capacity of municipalities and parishes, and will be based on the three principles of autonomy, increased subsidiarity and greater flexibility. To achieve this will require a series of changes including the decentralisation of new competencies, greater financial capacity (through income taxes and a larger redistributive grant from central government), reduced bureaucratisation, further rationalisation of local services, increased creation of municipal enterprises and the development of public-private partnerships, strengthened participation of local government in the spending of EU Structural Funds, and greater local government interaction with foreign organisations and local government structures.[37] For this type of major reform to progress certain basic changes in the operation of local government need to be implemented. First, the conditions under which local councillors operate need improvement, particularly in terms of higher salaries. Second, a larger support cabinet for the Mayor and councillors, and the possibility of defining the number of full time councillors, needs to be put in place.

In the short term, concern has focused upon evaluating the impacts of the legislative package adopted at the end of 1999. The extent to which this legislation permits progress in the priority policy areas of education and housing, as well as in strengthening local financial capabilities, reducing bureaucracy and ensuring effective participation in the Community Support

Framework for the 2000-2006 period, are the main issues under scrutiny. The local elections in 2001 seem set to be significant with regard to local government. The PSD announced proposals for a new model of local government in order to address the problems facing a rapidly changing Portuguese society, and the other parties responded with their own new policy proposals. The combination of profound economic and societal change together with the political attitudes already discussed, suggest that conditions are in place to facilitate the consolidation and development of the new forms of urban governance which have emerged over the last 10 years.

Acknowledgements

This chapter is based on research projects financed by JNICT (*Junta Nacional de Investigação Científica e Tecnológica*) and by FCT (*Fundação para a Ciência e Tecnologia*), developed at the CEG (*Centro de Estudos Geográficos*) University of Lisbon.

Notes

1   See INCM, Constituição da República Portuguesa (1997 revision).
2   For a more detailed discussion of administrative law see: Amaral (1993); Sousa (1995); Oliveira (1993); Caupers (1994).
3   Several proposals have been made in the past in order to allow non-party lists to participate in municipal elections but have always been rejected by Parliament. However, the 1997 constitutional revision did introduce the possibility of such a change. The Socialist Party presented to Parliament a new piece of legislation related to this issue in December 1999 and further progress on this is currently under discussion.
4   The Parliament Act on Local Government explicitly defines the competencies of both the executive and deliberative boards with respect to which body prepares, approves and executes the annual budget and the annual accounts, who passes certain rates and charges, loan contracts, etc..
5   Law 169/99 (18 September 1999) established the new regime of local government functioning and competencies, replacing law 100/84 (29 March 1984). According to the last Constitutional revision it is expected that the election to the executive board will change to a majority 'winner takes all' system.
6   Decret-law 116/84, 6 April 1984, with later amendments.
7   Only 4% (12) of Mayors are currently women.
8   The Local Finance Law 42/98, 6 August 1998 was approved solely with the votes of the Socialist Party.
9   The Municipal General Fund (FGM) is divided into three components (Mainland Portugal, Azores and Madeira) on the basis of criteria relating to population size (50%), number of municipalities (30%) and area (20%). Subsequently, municipalities

receive money on the basis of the following criteria: 5% for all municipalities; 35% proportionally related to population and tourist levels; 30% related to area and altitude; 15% related to number of parishes; 5% related to size of population under 15 years; and 10% proportional to personal income tax revenue (IRS) within the area.

10   The legal regime for this type of enterprise was established by Law 58/98, 18 August, 1998.

11   In 1982, the regionalisation process almost reached a positive conclusion. However the dissolution of the PSD/CDS coalition government brought this attempt to an end.

12   Law 56/91, 13 August, 1991.

13   Other factors which also appear to have influenced the result include: the unclear division of competencies between the state, regions and municipalities; the unclear meaning of the two questions included in the referendum; a fear that administrative regions would lead to the creation of regional parties; and the small size of certain regions (some of which were smaller than the municipality of Lisbon in population terms).

14   According to MEPAT (1998, p.23) there are 30 different sets of geographical divisions currently in use across different Central Government departments.

15   Law 44/91, 2 August 1991 created two Metropolitan Authorities (MA) in Lisbon (now with 19 municipalities) and Porto (with 9 municipalities). The PSD presented proposals for the creation of two new Metropolitan Authorities, in Aveiro and Leiria, but both were rejected by the Socialist Party. For further evaluation of the performance of the two MA's see Pereira and Silva (1999).

16   Law 172/99 (Municipal Associations. Law) and 175/99 (Parish Associations Law), 21 September, 1999.

17   For example in 1999 the national parliament discussed a range of proposals (to transfer new competencies to municipalities, create new metropolitan areas, reinforce municipal associations, develop municipal police forces, develop a new expropriation code etc.) which resulted in a new package of legislation towards the end of this year.

18   PCM, *Programa do XIV Governo* (XIV Government Programme, 1999), p. 138.

19   This is one of the policy areas included in the XIV Government Programme (1999-2003). The Socialist Government announced the creation of a Commission to review the reform of the state administration and to prepare for the 'de-concentration' of departments in January 2000. Other parties also support such reform. The CDS/PP have proposed the transfer of the Bank of Portugal from Lisbon to Oporto and the Constitutional Court from Lisbon to Coimbra and a reorganisation based on the pre-existing 18 *Distritos*. Even the Social-Democrat party has proposed this type of reorganisation, despite its reinforcement of centralisation tendencies whilst in office (1985-1995).

20   Law 58/98, 18 August 1998. Before this legislation at least two municipalities had already created municipal public enterprises through a "special" interpretation of the law.

21   Recent examples include the creation of municipal enterprises in Lisbon since 1994 (e.g. EMEL for car parking, GEBALIS for social housing estates management, EMERLIS for basic infrastructure management, EBHAL for the management of equipment in historical neighbourhoods, etc.), as well as in Cascais (for the management of social housing estates). In Nazaré a society for the implementation of a marina and related tourist activities has been created with the municipality providing 35% of the capital. The municipality of Braga is responsible for the creation

of three municipal enterprises (for the environment, urban public transport, and to run the municipal Expopark).

22 For example in the area of water policy, the ANMP in November 1998 concluded that the privatisation of water services was one means to achieving the objective of a better quality of life, as long as strategic control over the sector was maintained.

23 The failure of the Socialist Party to fulfil their electoral promise to double, in real value, the percentage of grants transferred to local government in four years, was a main focus of the PSD campaign during the 1999 national elections.

24 The ANMP wants municipalities to have the power to establish rates of municipal tax according to limits defined by Parliamentary law, alongside compensation for any local tax reductions adopted by central government. It also argues that there should be no exemptions of local taxes for central government departments (except in exceptional conditions), that there should be a periodic property revaluations, and that the possible introduction of new local fees with an environmental character, should be considered.

25 For instance in April 1999, the ANMP rejected proposed new competencies in the health and judicial sectors due to the lack of resources being provided to meet these new responsibilities.

26 Law 159/99, 14 September 1999, established the framework of new competencies to be transferred to municipalities.

27 Another means by which municipalities might be able to gain the "critical mass" required to deliver a wider range of services, could be via the formation of municipal associations specifically for this purpose.

28 Proposals for administrative modernisation include the publication of a Municipal Code, the introduction of ex-post verification by the Auditing Court for certain municipal decisions, the use of tacit approval in relation to central government decisions, the possibility of greater delegation of powers by local councillors to local technicians, and the introduction of analytical accountancy in order to improve municipal management. There are already signs that new practices, such as the use of external audits, are being introduced into current practice.

29 The purpose of a major conference held in May 1999 was to present and discuss best-practice examples in administrative modernisation within Portuguese municipalities. The DGAA (Central Government Department responsible for local government affairs) runs a programme which supports modernisation initiatives by municipalities and the AMDS (Municipal Association of Setúbal) runs a programme for municipal modernisation which has a reputation for promoting innovatory practice.

30 Local politicians from all parties are in favour of a metropolitan form of directly elected authority, including the Social-Democrats (Research Project Praxis XXI/ PCSH/P/GEO/50/96: Great Urban Areas).

31 The new planning system is set down in Law 48/98, 11 August, 1998.

32 Law 555/99 (16 December 1999) regulates the development process and Law 168/99 (18 September, 1999) approved the new Expropriation Code.

33 For example, in 1999 the municipality of Castelo Branco prepared together with the Ministry of Economy, a "*PAM – Plano de Acção para o Aproveitamento dos Recursos Endógenos e Gestão da Energia nos Municípios*".

34 The ANMP has agreed to accept a new competence for air quality but only given adequate financial support via EU Structural Funds.

35 See Ministry of Environment (1998), *Annual Report on the Quality of Water for Human Consumption.*

36 The best known example is the network of Business Information Centers (*Gabinetes de Apoio ao Empresário*) created and co-ordinated by AMDS (Setúbal Municipal Association) on the south-bank of the Lisbon Metropolitan Area.

37 One area of co-operation already well developed is with municipalities in the PALOPs (former Portuguese colonies in Africa) and, since 1999, with East Timor.

References

Amaral, D.F. (1993), *Curso de Direito Administrativo (vol.1)*, Livraria Almedina, Coimbra.

ANMP (Associação Nacional dos Municípios Portugueses) (several years) *Boletim da ANMP* ANMP, Lisbon.

ANMP (Associação Nacional dos Municípios Portugueses) (1998), *XI Congresso*, ANMP, Lisbon.

Barreto, A. (ed) (1998), *Regionalização: Sim ou Não?* Publ. D.Quixote, Lisbon.

Caetano, M., Barata, J., Esteves, M. and Pessoa, V. (1982), *Regionalização e Poder Local em Portugal*, IED, Lisbon.

Caupers, J. (1994), *A Administração Periférica do Estado*, Editorial Notícias, Lisbon.

INCM (Imprensa Nacional Casa da Moeda) (1997), *Constituição da República Portuguesa*, INCM, Lisbon.

INCM (Imprensa Nacional Casa da Moeda) *Diário da Assembleia da República*, INCM, Lisbon.

INCM (Imprensa Nacional Casa da Moeda) *Diário da República*, INCM, Lisbon.

INE (Instituto Nacional de Estatística) (several years), *Estatísticas das Administrações Públicas*, INE, Lisbon.

INE (Instituto Nacional de Estatística) (several years), *Estatísticas das Finanças Públicas*, INE, Lisbon.

MEPAT (Ministério do Equipamento, Planeamento e Administração do território) (1998), *Descentralização, Regionalização e Reforma Democrática do Estado*, Comissão de Apoio à Reestruturação do Equipamento e da Administração do Território, Lisbon.

Oliveira, A.C. (1993), *Direito das Autarquias Locais*, Coimbra Editora, Coimbra.

Oliveira, C. (ed) (1996), *História dos Municípios e do Poder Local*. Círculo de Leitores, Lisbon.

Oliveira, L.V. (1996), *A Regionalização*, Edições Asa, Porto.

Oliveira, L.V. (1997), *Novas Considerações sobre a Regionalização*. Edições Asa, Porto.

Pauleta, C. (1999), 'As Políticas Sociais das Freguesias na Área Metropolitana de Lisboa (1993-1997): uma Análise Geográfica.' *Dissertação de Mestrado em Geografia Humana e Planeamento Regional e Local*, Faculdade de Letras de Lisboa, Lisbon.

PCP (Partido Comunista Português) (1997), 'O Poder Local e as Eleições Autárquicas Resolução', *National Conference*, PCP, Lisbon.

Pereira, M. and Silva, C.N. (1999), 'Área Metropolitana de Lisboa: Balanço e Perspectivas', in *Actas do VIII Colóquio Ibérico*, September 1999, Lisbon.

Pereira, M. and Silva, C.N. (2000), 'As Grandes Áreas Urbanas. Contributos para a Definição de Alternativas ao Modelo Institucional Vigente', paper presented at *Seminário Internacional: Território e Administração*, February 2000, Lisbon.

Presidência do Conselho de Ministros (1999), *Programa do XIV Governo Constitucional*, Lisbon.

PS (Partido Socialista) (1995), *Programa Eleitoral do Governo e da Nova Maioria. Legislativas 95*, PS, Lisbon.

PS (Partido Socialista) (1999), *Programa Eleitoral do PS: Legislativas 99*, PS, Lisbon.

PSD (Partido Social Democrata) (2000), *Documento de Trabalho sobre as Autárquicas 2001*, PSD, Lisbon.

Reis, A. (ed) (1994), *Portugal. 20 Anos de Democracia*, Círculo de Leitores, Lisbon.

Santos, B.S. (ed) (1993) *Portugal: um Retrato Singular*. Edições Afrontamento, Porto.

Santos, J.A. (1985), *Regionalização. Um Processo Histórico*, Livros Horizonte, Lisbon.

Silva, C.N. (1995a), *Poder Local e Território. Análise Geográfica das Políticas Municipais, 1974-94*, Universidade de Lisboa, Lisbon. (unpublished).

Silva, C.N. (1995b), 'Autarquias Locais e Gestão do Território. Que Diferença Faz o Partido Político', *Finisterra*, 59-60, p.99-120.

Silva, C.N. (1996), 'Associações de Municípios e Regiões Administrativas', *Revista da Faculdade de Letras*, no.19-20, 5ª Série, pp.225-232.

Silva, C.N. (1998a), 'Local Finance in Portugal: Recent Proposals and Consequences for Urban Management', *Environment & Planning, C, Government and Policy*, vol.16, pp.411-421.

Silva, C.N (ed) (1998b), *Poder Local e Políticas Sociais em Portugal: Subsidiariedade e Parceria na Segurança Social (Relatório Final)*, CEG, Lisbon.

Silva, C.N. (1999), 'Local Government, Ethnicity and Social Exclusion in Portugal', in A.Khakee, P.Daest and H.Thomas (eds) *Urban Renewal, Ethnicity and Social Exclusion in Europe*, Ashgate, Aldershot, pp.126-147.

Sousa, M.R. (1995), *Lições de Direito Administrativo*, Lisbon.

STAPE (Secretariado Técnico para os Assuntos do Processo Eleitoral) (1987), *Caracterização dos Eleitos para as Autarquias Locais 1982*, STAPE, Lisbon.

STAPE (Secretariado Técnico para os Assuntos do Processo Eleitoral) (1993), *Caracterização dos Eleitos para as Autarquias Locais 1989*, STAPE, Lisbon.

STAPE (Secretariado Técnico para os Assuntos do Processo Eleitoral) (several years) *Local Elections Results*, STAPE, Lisbon.

Syrett, S. (1995), *Local Development: Restructuring, Locality and Economic Initiative in Portugal*, Avebury, Aldershot.

Syrett, S. and Silva, C.N. (1998), 'Local Government in an Era of Globalization: Evidence from the European Periphery', Paper presented at *Second European Urban and Regional Studies Conference*, 17-20 September, University of Durham.

Syrett, S. and Silva, C.N. (2001), 'Regional Development Agencies in Portugal: Recent Development and Future Challenges', *Regional Studies*, Vol.35, no.2, pp.174-180.

10 Challenge and Change: Prospects for the 21st Century

DAVID CORKILL

The naming of the new trans-Tagus bridge, opened in 1998, after the great navigator Vasco da Gama consciously made the link between achievements during the age of discoveries and the current era of European orientation. Portugal has established a stable democracy, from distinctly unpromising beginnings, and integration into Europe has triggered fast economic growth, offering the prospect that the economic gap separating Portugal from its European Union (EU) partners might yet be substantially reduced. The relationship with Europe has been the key to this transformation. Portugal's face is now firmly turned towards the continent whereas three decades ago this course was by no means certain. The intervening years have witnessed the abandonment of an imperial past, symbolically encapsulated when the last vestige of its overseas empire, Macao, was handed over to China in 1999, and attempts to graft a Lusophone heritage onto a European orientation.

Little more than a decade ago attaining the twin goals of political and economic stability appeared to be a remote prospect. Prior to 1987 no government had completed its allotted time in office. Since then, every administration has lasted the full four-year term. Such has been the resilience of recent governments that the minority Guterres administration (1996-2000) sustained embarrassing defeats in two referendums held in mid-term only to be re-elected by the voters. During the years of instability following the Revolution the presidency proved to be a vital cross-beam supporting the structure of Portuguese democracy. That there have been only three Presidents since 1976 has contributed greatly to the stability of the political system. Alongside a stable polity has come economic stability. After years of high inflation, a volatile exchange rate, and soaring interest rates, Portugal developed a stable currency, now firmly anchored in Euroland, growth rates regularly above those posted by most of its EU partners, and a narrowing of the gap in GDP per head with the European average.

Portugal's accession to the European Union in 1986 initiated a new phase of economic liberalisation. Although the beginnings of this process can be traced back to the 1960s, it is impossible to minimise the importance of the European commitment. The modernisers were able to argue that successful economic integration required major structural reforms and the establishment of a full-fledged market economy. An external stimulus was required because the centralised system and the large state sector were deeply rooted in the recent authoritarian and revolutionary past. Reform was made even more difficult because centralisation and state intervention were synonymous with the birth of democratic politics. Once initiated, an ambitious and extensive privatisation programme launched during Cavaco Silva's premiership (1985-95) served as the cornerstone for modernisation efforts. As Sousa's (Chapter Six) case study of the media and telecommunications sector demonstrates, this process led to a major redefinition and contraction of the role and influence of the state-owned sector. As a result, whereas in 1988 state-owned enterprises accounted for 19.4 per cent of GDP and 6.4 per cent of total employment, by 1997 it had been reduced to 5.8 per cent and 3.3 per cent respectively.

The reforms, which involved restructuring the financial sector and upgrading the transport infrastructure, were complemented by, and partly contingent upon, substantial financial transfers in the shape of Structural Funds emanating from Brussels. EU funding schemes made a significant contribution to the Portuguese economy, estimated as equivalent to 3 per cent of GDP in the late 1990s. The monies were directed towards investment in infrastructure improvement, workforce training programmes, and regional development projects. EU funds have been supplemented by the rising levels of foreign direct investment which, although varying year-on-year, contributed to the strong growth rates recorded in the late 1980s and early 1990s. When economic growth decelerated, Portugal benefited from joining the Exchange Rate Mechanism in 1992 and the pursuit of policies that allowed participation in Economic and Monetary Union (EMU) from January 1999.

Since joining the European Community the wealth differential separating Portugal from its EU partners has narrowed considerably. Eurostat calculated that per capita GDP, based on Purchasing Power Parities, rose from 55 per cent of the EU average in the early 1980s to over 70 per cent by the late 1990s. Indeed, apart from Ireland's 'Celtic tiger' economy, Portugal registered the best performance among the cohesion group countries (Ireland, Greece, Portugal and Spain) and achieved the greatest improvement in living standards, relegating Greece to the position of poorest member state. Generally speaking, Portugal has usually grown fastest during high growth periods in the European economy. However, it has lagged behind during economic slowdowns. This sensitivity to the health of the European economy is important given the need to grow at a faster rate than EU average to close the gap. The economic challenge facing Portugal is to maintain the rhythm of growth during recessionary downturns in order to ensure that the gap continues to close.

It is a measure of the progress achieved over the previous two decades that by the end of the twentieth century Portugal faced a new set of challenges. Some were shared in common with other mature democracies (the 'participation crisis', immigration, the environment), or stemmed from the EU integration process (enlargement, growth and convergence, regional inequities), while others were structural problems that had been on the agenda for some time but still remained to be tackled (bureaucracy).

The Challenges of Maturity

As noted throughout this book, building a stable democracy since the mid-1970s ranks as a major achievement for a country with little or no democratic political tradition and an underdeveloped civil society. However the undoubted success story of the past three decades should not obscure the very real problems and challenges currently facing Portugal's democracy. As Robinson points out (Chapter Eight) Portugal is now a fully-fledged European democracy and the difficulties it faces are shared in common with its EU partners. Many modern democracies have been affected by public disenchantment with politics and politicians. Concern has been expressed that apathy may corrode the foundations of the political system.

In Portugal there is evidence of low and falling membership of political parties and trade unions, together with survey data revealing consistently low public esteem for institutions such as the National Assembly, political parties, the civil service, the Catholic Church and the armed forces. Generally speaking, politicians continue to be held in poor regard, perceived as motivated solely by electoral advantage and the spoils of office. There exists a widely held view that the stagnant political elite requires an injection of new blood in order to extend access in a polity still dominated by patrimony and patronage (Magone, 1997). An increasing number of voters, alienated by politics and unmotivated by the political leaders, have opted to abstain rather than go to the polls. The 2001 Presidential campaign provided a case in point, giving further ammunition to those who argue that Portuguese civil society is experiencing a 'participation crisis', a trend evident in successive elections at European, national and local levels, where fewer and fewer voters have chosen to exercise their democratic rights.

What lies at the root of this political apathy? At the outset it has to be recognised that the spread of this political malaise is not exclusive to Portugal. Electoral participation has declined in every Western democracy. Low voter turnout may reflect a growing indifference to politics, but disengagement should not necessarily be interpreted as outright rejection. When the incumbent President Jorge Sampaio was re-elected to a second term in the January 2001 Presidential election, abstention reached a historic high at 49.1 per cent of the registered voters. Most commentators attributed voter apathy to the absence of burning issues and the strong likelihood that Sampaio would win comfortably. The main opposition candidate, Ferreira do Amaral, blamed the growing gap between governors and the governed and warned of 'symptoms of fragility' affecting Portugal's democracy. Sampaio made the obligatory promises to 'mobilise the citizens' and overcome the inertia and resignation, but the general feeling was that little could have been done when the election result was never in any doubt. Clearly it was hardly an exciting prospect to vote for stability and continuity. Nor did it help that the outdated electoral register still contained the names of some half million deceased voters.

It is equally interesting to note that in Portugal's semi-presidential system, the presidency has effectively become a ten-year job. The three incumbents since the mid-1970s, Eanes, Soares and Sampaio, have all served the maximum two terms permitted by the constitution. In 2001 the voters denied the presidency to the Right and the Socialists retained control at presidential, parliamentary and municipal levels. Perhaps just as worrying for the long-term health of democracy was the one-sidedness of the contest. Concern was expressed that the opposition, led by the Social Democrats (PSD), failed to offer an alternative and fielded only a second-rank politician as its standard-bearer. It remains to be seen whether electoral reform, and changes to the party list system in particular, will succeed in revitalising democracy and boosting popular participation. In fact it may be that the most vibrant and participatory politics in Portugal are now to be found outside of the formal political system. Single issue interest group politics, perhaps most visible in the environmental arena (see Chapter Seven) but also evident in the local boycotts of the 2001 Presidential elections (Diário de Notícias, 2001), are growing, as are an array of new urban movements.

Portugal's transformation from a labour exporter to a net importer must be regarded as a major social development in late twentieth century Portugal (see Chapter Five). Increasing prosperity and near full employment stoked the demand for immigrant labour and transformed Portugal into a receiver society and a 'new immigration centre'. Major current infrastructure projects, including the second Lisbon airport at Ota and the stadia for the Euro 2004 football tournament, ensure that the demand for labour is set to remain high. Although, comparatively speaking, the numbers of legal and illegal foreigners remain relatively small, the immigration issue has been forced higher up the political agenda. A number of factors explain this development. Internally, Portugal's low unemployment rate ensured tight labour market conditions and immigration was a response to a labour-short economy. Externally, pressure came from the EU to toughen entry controls, challenging the cherished traditions of openness and tolerance for which the Portuguese have been justly proud (Corkill and Eaton, 1999).

In the past, when the overwhelming majority of immigrants originated in Portugal's former colonial empire, cultural and linguistic links facilitated integration. However, as Eaton points out (Chapter Five), the changing composition of the immigrant flows, as East Europeans from Russia, Romania, Moldova and the Ukraine have joined Asians and immigrants from the traditional source countries (Cape Verde. Guinea-Bissau, Angola and Brazil), poses new challenges. The historic ability to absorb large numbers of incomers - demonstrated most notably in the mid-1970s when up to three-quarters of a million refugees fleeing the chaos in the war-torn former African colonies flooded into the country - stood Portugal in good stead. However, there have been some isolated racist incidents and a growing concern revealed by the findings of public opinion surveys, linking immigration to crime and growing insecurity. Faced with an influx of immigrant workers the official response has been to issue temporary permits, valid for up to five years, in an attempt to meet the labour shortfall. This short-term solution, narrowly based on labour market requirements, failed to address the challenges of multiculturalism and integration posed by issues of 'ghettoisation' and poor quality housing, labour exploitation by unscrupulous employers, and social inclusion.

In the frantic push to sustain a high rhythm of economic growth, concern for the environment can easily be sidelined (see Pereira de Silva, Chapter Seven). It is a reflection of the continued weakness of civil society that the Portuguese have belatedly recognised that development brings social and ecological costs. Only the more reflective, like the Nobel Prize for literature novelist José Saramago, paused to question the precedence given to the economy to the detriment of the environment, education and culture (Brassloff, 1998). Much development, particularly during the 1980s, was unregulated and unplanned. More recently, the clash between economic development and environmental conservation has been vividly illustrated by the Alqueva dam project. The scheme to create what will become the largest artificial lake in Europe, close to the Spanish border in the Alentejo region, ignited much controversy during its long gestation and construction. The rationale driving the project was laudable enough; to provide irrigation for the production of fruit, vegetables, flowers and other crops in a declining region suffering from unemployment and high out-migration. However, the construction work threatened the unique eco-system in a region noted for its wildlife, including many rare and endangered species, and cork woodlands. The European-funded development required that one million trees be felled in order to make way for the lake and develop the associated tourist facilities (Gonçalves, 2001).

The European Challenge

Portugal could be one of the chief sufferers from the EU enlargement process (potentially twelve applicants could be added) largely because the new members, also low cost labour nations, will compete for public and private investment. However, wage flexibility in the labour market, Portugal's traditional mechanism for coping with external competition, could still help to offset the threat, as could the industriousness of the workforce, reflected in the prevalence of multiple employment at double the EU average. A more significant threat may in fact derive from the attempt by the larger member states (France, Germany, Italy, Spain and the UK) to reform EU institutions in their favour. The Nice summit underscored the danger that the next stage in the integration process will curtail the powers of smaller states like Portugal, who may find that they are permanently marginalised from the centres of power.

Despite the considerable benefits that Portugal has derived from participation with its European partners, there are dangers inherent in closer European integration. Unless corrective measures are taken, social disparities could widen, regional disequilibria worsen, and exclusion grow, while the environment will suffer further destruction in pursuit of growth and energy sources. Economic as well as political sovereignty might be surrendered in pursuit of growth and convergence, reducing a once great colonial power to a peripheral outpost of other economic empires. However, the counter-argument is that a high price might be paid in order to retain national sovereignty in a globalised world and that supranational structures at least afford some protection from external competition.

Economic growth has not, as Syrett (Chapter Three) confirms, reduced regional inequalities. Although substantial progress has been made towards closing the per capita income gap with its EU partners, wealth is still unevenly distributed. Indeed, the evidence suggested that fast economic growth contributed to a widening, rather than a narrowing, of the differentials between the rich and poor regions. According to data from the European Commission, the trend towards worsening regional disparities within countries during the pre-1996 period has stopped, although it is too early to say that it has been permanently reversed. It is salutary to learn that a reduction in the disparities between the interior and the coast can probably be explained by migration to the richer regions rather than by growth in the poorer ones (Público, 2001a). Such processes of internal migration have fuelled rapid urbanisation and produced a range of new urban problems (e.g. drugs, crime, social exclusion) often focused in pockets of intense deprivation within the expanding major conurbations. The gap between the richest region, Lisbon and the Tagus Valley and the poorest mainland region, the Alentejo, remains a glaring one. Nor has the buoyant tourist industry helped to bridge the regional divide. The evidence provided by Williams (Chapter Four) illustrates that tourism contributed to reinforcing regional inequalities. Nevertheless, progress has been made in infrastructure modernisation, particularly the road network and rail links. In this regard, the Lisbon area has benefited the most and can boast an impressive array of public works projects, including the metro extension, the second Tagus bridge and the Expo-98 site.

While it is indisputably the case that closer integration with Europe has brought significant advantages, there exists the danger that some of the challenges still to be faced have been minimised or unwisely ignored. Despite the considerable economic progress made over the last decade and a half, many Portuguese still do not have a standard of living comparable to that enjoyed by the populations of their EU partners. According to Eurostat's definition of poverty, Portugal has the largest number of poor (measured as income less than 60 per cent of the average in each country) in the EU. A quarter of the population is classified as poor - a proportion that would be even higher if calculations were made using the EU average. To close the gap and reduce existing poverty will require sustained high levels of economic growth and a significant strengthening of welfare services. Worryingly, after 2006 fast growth may have to be achieved without the generous assistance provided by European funds and the buoyant levels of foreign direct investment enjoyed during the 1980s and 1990s. In such circumstances it would take up to twenty years for Portugal to reach average EU wealth levels, even under the most optimistic predictions (Wise, 2000).

Structural Challenges

Rapid social and economic change has not been matched by parallel modernisation of Portugal's public administration. Reform of the unwieldy and change-resistant bureaucracy has long been identified as a necessary structural reform that successive governments have paid lip-service to, but failed to tackle. One of the signature features of the post-1986 era has been the retreat of the state, which has a much-reduced economic role following the privatisation programme and the powers ceded to Brussels. Nevertheless, Portugal is still largely a centralised state and limited progress has been made towards decentralisation. As Silva argues (Chapter Nine), local government still operates within a system best described as one of 'hierarchical dependency' in which central government has control over the purse-strings. There is a general consensus that civil service costs need to be trimmed and brought in line with the general change in the relationship between state and society. However, the political will has not existed to carry out the necessary reforms, and public expenditure has continued to rise, growing by eight per cent in the 1990s, compared to a four per cent fall across the EU (Público, 2001b).

Low productivity has been identified as one of the principal constraints on catching-up and a contributor to economic growth deceleration. Inside the eurozone the Portuguese authorities can no longer resort to exchange rate manipulation in order to boost exports and attract foreign investors. Successive governments have been well aware that the priority must be to improve the quality, design and image of Portuguese products, upgrade skills to meet future labour demand by investing in education and training, and encourage the use of information technology through broader access to the internet. Above all, there is a recognition that a psychological change is needed. This can only be achieved by investing in management, reforming business practices, adding value to manufacturing and services, and improving output per head (The Economist, 2000).

Final Thoughts

On the cusp of the new century Portugal presented a mixed picture. There was a very real feeling that the country had reached a crossroads. The rate and scale of change had been impressive, yet much remained to be done to close the gap separating Portugal from the advanced European economies. In many respects the 'window of opportunity' afforded Portugal after its entry into the European Community was beginning to close and by 2006 might shut for good, unless efforts to sustain continued EU funding after that date bear fruit. The frenzied development and modernisation did much to transform the economy and society, but some areas remained strangely untouched. Aspects that marked the country out as advanced and pioneering (electronic motorway tolls, high mobile phone ownership rates, the Multibanco automatic banking system, e-commerce in a growing internet market), co-existed with negative features from a society still exasperatingly hide-bound by bureaucratic red-tape, inefficient public services, and patrimonial politics.

The change in the national mood from optimism to introspection was reflected during 2001. It was then that the euphoria and optimism generated by the successful world exhibition, the choice of Oporto as European Capital of Culture, and the award of the European football championship in 2004 was punctured by two transport disasters. The collapse of a road bridge, killing more than 50 people, and, in another incident, a coach accident in central Portugal, claiming 15 lives, underlined just how antiquated much of the infrastructure still remained. A mood of introspection swept across the country following the Entre-os-Rios bridge collapse. Many asked whether progress should be measured solely in terms of cement and asphalt. For every new motorway and bridge there were ten roads and crossings requiring urgent attention. Further doubts remain over whether future economic growth will be able to generate the wealth necessary to maintain and improve the array of new infrastructures developed over the last 20 years. In this respect, the EU's largesse had not proved to be a magic panacea any more than the Brazilian gold that poured into the country centuries earlier.

The focus in this volume has been on political and economic change. However social change has been just as marked as the economic and political transformation. The switch from a country of emigration to a receptor society, the shift towards urban living, the creeping secularisation in a once deeply Catholic country, the changing role of the family and the transformation in women's lives - particularly their participation in the labour market - have all impacted deeply on Portuguese society. In some respects the country has been changed out of all recognition. Certainly there has been a major change in how the Portuguese see themselves and their role in the world. The difficult switch has been made from an imperial to a European vocation. In achieving this a new identity has been forged; one that combines elements of the past with a more self-assured, open and forward looking outlook.

References

Brassloff, A. (1998), 'Portugal Society and its Values in 'the Cavaco Years'', *Portuguese Studies Review*, vol.7, no.1, pp.70-83.

Corkill, D. and Eaton, M. (1999), 'Multicultural Insertions in a Small Economy: Portugal's Immigrant Communities', in M. Baldwin-Edwards and J. Arango (eds), *Immigrants and the Informal Economy in Southern Europe*, Frank Cass, London, pp.149-168.

Diário de Notícias (2001), 'Dia de eleições com muitos boicotes', 15 January.

The Economist (2000), 'Half-way there', Survey, 2 December.

Gallagher, T. (1999), 'Unconvinced by Europe of the Regions: The 1998 Regionalization Referendum in Portugal', *South European Society & Politics*, vol.4, no.1, pp.132-148.

Gonçalves, E. (2001), 'Death of a valley', *The Guardian*, 21 February.

Magone, J. (1997), *European Portugal. The Difficult Road to Sustainable Democracy*, Macmillan, Basingstoke.

Público (2001a), '25 anos para chegar à média europeia', 26 January.

Público (2001b), 'Olhar para frente', 7 April.

Wise, P. (2000), 'Nation in a hurry wants to travel faster', *Financial Times*, Annual Country Report, October 25.

Index

231

Printed and bound by CPI Group (UK) Ltd, Croydon, CR0 4YY

22/10/2024

01777626-0001